The Ones Left Behind

A Journey of Love and Loss
of Faith and of Hope

Endorsements

Mike's kind heart, intellectual sharpness and passion for Jesus draw you into his friendship and his book. Mike speaks eloquently from his own personal experience in his journey navigating profound grief following the early death of his beloved wife Charmaine. He shares how his faith in Jesus brings comfort, joy and hope in a difficult time with lessons for us all in the journey through life. Mike is an inspiration!

Andy Shepherd, Finance Director

Mike's book, like Mike himself, is thoughtful, faith-building and gently provocative. His articulate generosity of spirit shine through, in this personal story that invites its readers to go further and deeper into the mystery of God's goodness, despite the all-too-real pain of life and loss.

Amy Hailwood, Theatre Director

Michael is a great thinker, and every time I meet with him he shares new ways of looking at things. Now he has brought several key themes together in The Ones Left Behind. Thank you Michael!

Martin Walli, Businessman

It may also be a separation and not death that makes you feel like the one left behind. Either way the pain and loss are very real. The hard questions in these times are real too and we seek real answers. Michael's own journey led him to pose such existential questions. Although he walks through them with his readers in ways that to some Christians may occasionally seem heretical, he is always true to Jesus and to the Word. Non-believers may find hope and solace along the road he takes to God through Science. It's not a quick read but may be the right read for many.

Andrew Kampouris, Businessman

In Mike's first book 'The One Left Behind' I was moved by his own story of love and loss. In 'The Ones Left Behind' he associates too with the grief of others whose loved ones went on ahead leaving them feeling left behind. He addresses questions everyone has such as 'Where are they now? Will I ever see them again?' He matches scientific facts about the universe together with Biblical facts in unique ways that strengthen faith and bring hope to those who have loved and feel their loss.

Rosie Park, Architect

Michael has written a must-read book for anyone who has ever questioned how science and faith can co-exist and still be true in the face of life-changing loss. His genuine warmth and curiosity guide the reader through thinking differently about long-held beliefs.

Julie Tushingham - Accredited Business Coach

The Ones Left Behind

A Journey of Love and Loss of Faith and of Hope

Michael Morris

Illustrations by Tom McCrorie

Publications

COPYRIGHT

Copyright ©Michael Morris July 2024.

Published: July 2024 Ladey Adey Publications, Ancaster, Lincolnshire UK.

Michael Morris has asserted his right to be identified as the author of this Work in accordance with the Copyright, Designs and Patents Act 1988.

ISBN: 978-1-913579-66-1 (Paperback).

ISBN: 978-1-913579-67-8 (E-Publication).

All rights reserved. No part of this publication may be reproduced, stored in a retrieval system, or transmitted in any form or by any means - for example, electronic, photocopy, recording - without the prior written permission of the publisher. The only exception is brief quotations in printed reviews.

British Library Cataloguing-in-Publication Data.
A catalogue record for this book is available from The British Library.

Cover Design by Abbirose Adey, of Ladey Adey Publications.

Cover Image created by Tom McCrorie

Neither the author nor the publisher can be held responsible for any loss, claim or damage arising out of the use, or misuse of the suggestions made, the failure to take business, financial or legal advice or for any material on third party websites.

The author and publisher has made every effort to ensure the external websites included in this book are correct and up to date at the time of going to press. The author and publisher are not responsible for the content, quality or continuing accessibility of the sites.

Copyright disclaimer: Content provided is independent and does not represent publisher's views, opinions, or endorsements in any manner.

Bible References: all taken from New International Version (NIV) unless otherwise stated.

If you have enjoyed this book please give a review on Amazon® for Michael.

If you wish to discuss any of the theories and thoughts raised in this book with Michael, please visit: micharm@theonesleftbehind.co.uk. Always remember to consider others and be kind.

Dedication

Dedicated to Charmaine,
who wrote in my 21st birthday card the day after hers

'Time and again,
however well we know the landscape of love,
and the little churchyard with the lamenting names
and the frightfully, silent ravine, wherein in all others end:
time, and, again, we go out two together,
under the old trees, lie down again and again
between the flowers, face-to-face with the sky.'
Rainer Maria Rilke

Dedicated also to all those who're feeling left behind

We shall not cease from exploration
And the end of all our exploring
Will be to arrive where we started
And know the place for the first time
T.S Elliott - Little Giddings

THE ONES LEFT BEHIND

Contents

Foreword - Ladey Adey ix
Preface xi
Introduction 1
1. Science as Reality 7
2. Jesus and Storms 29
3. Again a Bride 37
4. The First Storm 49
5. The Second Storm 61
6. The One Who Went Ahead 63
7. And Now? 75
8. The Third Storm 85
9. Conclusions So far 89
10. None Left Behind 93
11. The Power of God's Love 115
12. The Kingdom of God is Within Us - Scientific and Biblical Facts 121
13. Where did all this come from anyway? 157
14. What are we all really made of? 171
15. From Lowly Bodies to Glorious Bodies 203
16. When and How Will This Be? 209
17. God is Light 225
18. The Devil Was Defeated at the Cross - but How? .. 233
Conclusion 239
Epilogue: The Way We Were 241
Glossary Of Terms 243
References and Endnotes 255

Acknowledgements	263
About the Author	265
Index	267
Notes	269

Foreword

Foreword - Ladey Adey

Printing is the ultimate gift of God and the greatest one.
Martin Luther King

We receive many gifts from God - they include being able to think and having freewill. As Proverbs 23:7 tells us, 'For as he thinks in his heart, so is he or a less complicated way of understanding it is: As you think so shall you be!' It is up to us what and how we think and then what actions we take on them. We are not born to be forced in our thinking and ultimately have the choice of what we believe, how we behave and the decisions we make. There may be some grey matter in between but I shall leave this for the psychologists to deliberate.

The beauty of this work by Michael Morris is he is giving suggestions to help us to think beyond our habits and possible doctrines. He is suggesting that God has much more and has created much more if we take the time to look under the surface a little. It is more than what religious leaders may tell us and more than what scientists have discovered. In fact, the more of a puzzle it is the more God has the true answer but it not just given - we have to investigate and make our own discoveries.

It never fails to amaze me how we are taken on our own unique journey of life, with its ups and downs, twists and turns but at each point there is learning to be had and ways for us to keep our minds open as to what we may find.

In the same way, we cannot take anything for granted, our beliefs, our relationships or our health. Each can change quickly, some changes we can instigate, some are beyond our control, yet even these we can choose how to react to them! Either way, Michael is a champion of thinking deeper, wider and higher. He combines the human condition with learning from the scientific element and finds both in the Bible! Suggesting God knows what he is doing!

Referencing his own Bereavement journey with the loss of his beloved wife, Charmaine, Michael leads us to believe, loss can be faced without sorrow (sadness yes) as we have hope knowing that when Jesus died and came back to life again, we can be secure in knowing when Jesus returns, He will bring back with Him all the Christians who have died. (paraphrased from 1 Thessalonians 4:13-18).

Enjoy *The Ones Left Behind* and put on your thinking caps while you do, be challenged and see if you agree or can find where scientific finds and biblical elements align. Finally, be reminded by Michael of the saying, "It's not true because it's in the Bible, it's in the Bible because it is true."

Ladey Adey (Sept 23)

Preface

"All marriages are royal marriages."
The Bishop of London

We were together for nearly 50 years. My wife was more than enough, the love of my life, and my truest friend. She went on ahead, and now I'm the one left behind.

I met Charmaine on our first day at university. There she was with her back to me, standing on the dance floor. I touched her shoulder, she turned, there I was. I asked her for a dance and there began the rest of our lives. Only later, did I realise how much she was to me, echoed from the words of the song by Roberta Flack, *"The first time ever I saw her face, I saw the sunrise in her eyes. The moon and the stars were her gifts to me."* Over the years we had each become for the other, while hardly knowing it, quite simply our meaning in life, our reason to live. I once quoted Boris Pasternak in a letter to Charmaine, *"You came in with a chair, reached up to take my life, as from a shelf, and blew away the dust."*

To paraphrase the scripture, but to retain the spirit of its meaning, in each other we *"lived and moved and had our being."* (Acts 17:28)

Charmaine had just been moved into the home-care hospital bed. It was delivered the same evening, and I'd left her to take some rest after several sleepless nights. The bed was empty when I came back downstairs in the early hours. When I returned from looking for her in the kitchen, there she was lying beside the bed. I touched her shoulder, but this time she did not turn. Our lives together had ended. She'd rested her head on her right arm, making herself comfortable, as if thinking, *"I'll just wait for Mike to come back."* My love's left arm offered no resistance beyond its own weight as I reached over and lifted it to wake her. It felt strangely heavy to my hand. There was no sign whatsoever of the vivacious, outgoing, caring life I'd known. This time there was no familiar waking from sleep. This was the moment when I knew I had been left behind.

Charmaine lived with cancer for eight years and it challenged our faith. Losing her challenged mine even further, but Jesus came ever closer. Charmaine, towards the end of her life, wanted God to somehow have *"the greater glory"*, as she put it, whether she remained with me or went to be with Jesus.

A profound insight later helped me understand what happened that night. Jesus had been there. He'd come back for her, for his bride just as he promised he would. They'd only just left - together, hand in hand.

He then took me back to how well Charmaine and I had loved. We had always remained faithful to our first love for each other. When we met, we were far from knowing God at the time, but looking back we saw He knew us, was with us and for us even then. Before knowing why, we would often remark on how our love seemed to be so much bigger than what each of us contributed to it, as if there was still more love coming in from somewhere beyond us both. This can be a far from unique experience for so many for so long, but it was totally unique and extraordinary for us.

I know now how God somehow added His unconditional love to our own. We both felt Jesus's love for His bride(s) to be in addition to our own love for each other. I believe marriage models our relationship with Jesus. (Ephesians 5:21-33)

As the Bishop of London put it at William's and Kate's wedding, "All marriages are royal marriages."

Later that year I was re-baptised in the Jordan. Jesus spoke so kindly but so firmly to me of Charmaine having become His bride. He then healed my heart in a kind of rebirth into life as I will recount later.

For those who've lost a loved one, feeling *left behind* happens again every morning. They too may well wonder as I did, *"Where are they now? Will I ever see them again? Why would God ever care about me anyway - if He's even there?"* - and more.

Someone with experience once asked me, *"Is it well with your soul?"*[1] This led me to the song, which put words to what the Lord wanted me first of all to know and to do.

> *Far be it from me to not believe,*
> *even when my eyes can't see?*
> *Through it all it is well*
> *Through it all my eyes are on you*
> *It is well with my soul.*
> *Let go my soul and trust in Him*
> *It is well with me.*[1]

I still believe, but it was not always well with me, nor with my soul. Sometimes, you know, it was not well at all! There was work to be done, but here was a starting point. Many more songs and Bible verses have followed since then, as the Holy Spirit brought them to mind.

This book is not about how to do it, how to cope with grief and bereavement. It's not even about how to overcome loss. It's not about how to answer questions like, *Why?* and *Why me?* to which there can never be truly satisfactory answers. *Why?* may even be the wrong question. It's more about growing a relationship with God in ways which will not be possible in heaven. Although we are headed to a glorious heaven, bound to live on in eternity, here we have choices. Meanwhile, there is a unique purpose for us being *bound* here temporarily on earth. There are things we can learn, choices we need to make, and unique praises we can give under the pressure of earthly circumstances like these. They can only happen here, in this sliver of time between eternity past and eternity future.

The teaching on God not being the cause of sickness and untimely death, is something I *get*, I *understand* it. If we are in any doubt, then we can say with certainty, Jesus is perfect theology. He is *"the same today, yesterday and forever."* (Hebrews 8:28). All who came to Him were healed. (Matthew 4:24) I cannot afford ever to forget we are in control of our lives, we have the freedom of choice which Jesus died to defend. But He is in charge. He is Good all the time. He is also Wise.

This is not to lay claim to this being a handbook on how to deal with loss. Rather, it's a personal testimony to God's goodness, his love and the power of his love. It's more about *How?* than about *Why?* and certainly not, *Why me?* It's about how God demonstrated His goodness and kindness in the midst of circumstances, every aspect of which seemed to deny His love. But it's not just about me. In most close relationships, including our own parents', one will pass, go on ahead and one will be left behind. How are we especially as Christians, to navigate this? At times such as this, even our faith is challenged, perhaps as never before. Times such as this can grow and deepen faith as never before, but they also put it at risk of being destroyed.

This book is not for everyone, and I don't claim to have all the right answers, but it will hopefully address some of the right questions.

This is also the story of how I've grown to live without Charmaine, not by moving on but more by continuing to move forward. It's a testimony to how God's goodness is helping me do this with love and purpose. God is Good all the time... He is also Wise - he knows what he is doing.

In times of grief, we might find ourselves asking the kind of questions we never did before, pushing against the boundaries of credibility with other church members, friends and family - or alone with ourselves.

So, we need to pray to the Father,

> *"Out of his glorious riches he may strengthen us with power through his Spirit in our inner being, so that Christ may dwell in our hearts through faith. And pray that we, being rooted and established in love, may have power, together with all the Lord's holy people, to grasp how wide and long and high and deep is the love of Christ, and to know this love that surpasses knowledge—that you may be filled to the measure of all the fullness of God." (Ephesians 3:16-19)*

I'm sure many of us are familiar with these words. This is a testimony to how God provides for us, as the ones left behind. He even provides this power we need to grasp His love and to know His person, awake and in real time - when we ask, and sometimes in His kindness even when we don't. At the end of the day, it's what Jesus paid for us to have.

Even in Charmaine's passing I can relate to Jesus's reflection, *"the kind of death by which Peter would glorify God."* (John 21:19 NIV) Although in completely different circumstances, perhaps there really is a *"death that will glorify God."*

THE ONES LEFT BEHIND

Jesus shows us step by step, day by day, how we are to weather these storms of life, which grow in stages, and through them to know him better and to love him more.

Michael Morris

Introduction

*There are more things in Heaven and Earth,
Horatio, than are dreamt in your philosophy.*
Hamlet (Act 1 Scene 5)

This is my story of how God deepened my faith and put the science described here as an objective, hard back stop of faith to my own subjective reality. The cancer died, but Charmaine lives.

We are essentially body, soul, and spirit. (1 Thessalonians 5:23) Body and soul are physical, natural and temporary. We dwell in a tiny sliver of time and inhabit a tiny part of an all but infinite visible universe, itself a small part of all which really is. Yet our spirit is eternal, *"super-natural."* It came into existence before the body and soul and it lives on after they have, well, melted away. *"God is spirit."* (John 4:24) and I believe we too are spirit, *"supernatural beings inhabiting natural bodies."*

In a way, it's where the rubber hits the road. God, salvation, forgiveness, eternal life and everything else we read about in the Bible are either all true - or all false. The choice is ours. My heart wanted to believe, my mind wanted to understand, but my soul cried out, and still it gently weeps. Praise God, He does not leave us alone in making this almost daily choice. All this can only be reconciled where there is the peace God gives which *"transcends all understanding to guard our hearts and our minds."* (Philippians 4:7) Fortunately, we are not required to understand it.

I am persuaded, it's our daily choice to believe, or not. This choice is made in faith through, and not just despite, circumstances. This faith choice is at the heart of the *"life and life to the full"* (John 10:10) Jesus came here to give us. All the things we'd rather not have happen are included, as well as those we would. How fortunate we are when we allow Jesus surely to be with us always, *"to the very last minute"* (Matthew 28:20) to help us deal with these earthly choices, and way beyond.

Which takes us directly to my story of exploration of an all-or-nothing boundary of choice. There's always the choice between faith and doubt *as presented* by the circumstances which God allows yet manages, *"Lord I believe; help me with my unbelief."* (Mark 9:24)

Science alone has no answers to such questions of the heart, claiming there is no evidence for the existence of God. Yet, Voila! Here we are, as human beings, still forcefully pondering the answers to these very questions. We today are the net outcome of a long process - a 13.8-billion-year overture after the 'Big Bang' creation of the universe. Through the fulness of this very experience of grief and bereavement, I discovered a broad and powerful correlation between the science of the fullness of this process and the Bible. For myself it is an objective "hard backstop" of faith to my own subjective

INTRODUCTION

reality. So this we will explore in the coming pages - with a light touch.

A common myth of Science is, '*God does not exist because we have evolution.*' A common myth of Religion is, '*Evolution does not exist because we have God.*'

The extraordinary fact of the existence of our world as we know it, and our own very existence in it, begs the question, '*What after all would be sufficiently convincing scientific evidence of the existence of God?*'

Whatever form this scientific evidence might take it would have to be significant enough to remove all doubt from our minds. Yet, such evidence would therefore remove all human freedom of choice. Would we not be returned by default to a mediaeval state of total subservience imposed this time by Science instead of by the religion of the day?

Which is not to say the Bible has a monopoly on truth and morality. But without the freedom to choose to do wrong as well as do right we as individuals would lose one of the few factors differentiating us from all other living things; the freedom to make moral choices, to get them wrong sometimes and learn from them to do better.

Where Science looks to understand God through debate without referring to the Bible it's like looking to understand mathematics without referring to equations. I look to relate Science to the Bible.

One big question often posed is, how can God be three things - at the same time? I like to use Science to *backstop* this question of faith. I believe God the Father, Jesus and the Holy Spirit are all 100% God, while Jesus was and is also 100% human. For example, Ice, Water and Steam are all H_2O - at the same time. They manifest in totally different forms with totally different qualities, but each is totally consistent with the other. In a quite extraordinary way they each have the same identical molecular composition of two Hydrogen atoms bonded to one Oxygen atom - everywhere in the universe, always. Likewise then God is God - everywhere, always, however and wherever He may be, as Father, Son and Holy Spirit.

Another huge debate links back to before the beginning of time itself; before the 'Big Bang' 13.8 billion years ago, the universe was not - until it was. However, the universe could not have just created itself.

As we look to the origins of Life itself, we find a similar void to the one around the origin of the Universe. Before its own 'Big Birth' 4 billion years ago, Life was not - until it was. Similarly, Life could not have just created itself or evolved itself.

What we do know is Genesis 1 and 2 do not use Einstein's description of the origin of 'Spacetime'. But, *"God knows the end from the beginning."* (Isaiah 46:10) He knew of course, Newton, Rutherford, Hubble, Einstein and many others would come along later to uncover some of the details for us.

The Bible was written at least 4,000 years ago in an accessible, non-scientific allegorical form. I believe it's like God telling people of the day (and most of us even now), *"Don't bother your head too much about all the details. The main thing is this. I was there already, even before the beginning of time. Then I just said so, and time and everything else began."*

Instead, I think God wants us to grasp the really important themes of Genesis 1 and 2, like love, forgiveness, identity, destiny and Jesus and take them into our daily lives and beyond.

As I see the Bible either it's all true, or none of it's true.

It's not necessarily true as in everything happened literally in just the way it's described in Genesis 1 and 2. But it's all true as in God chose to put it this way according to his wisdom,

INTRODUCTION

not ours. Which means things are not true because they're in the Bible; they're in the Bible because they're true.

To summarise, about both the Universe and about Life, *"In the beginning God said..."* (Genesis 1, John 1:1-4) and there they were. Science tells us the how, the Bible tells us the who, and the why.

My purpose for writing this book is for God to use it like *"loaves and fishes"* (Matthew 14:17-20) for his *'greater glory'* in at least one way if not many, with at least one person if not many. It might enable you to come to know him better for who he is and for who he says you are - so beloved by him he sent his only Son - God Himself to tell us.

I refer quite frequently to the Bible. Thanks to Bible Gateway I can find what comes to mind as I write, not knowing very well how else to find where it comes from in the Bible. Some of you will be very familiar with scripture, some not. I'd like to encourage you at least to read these best-of extracts through on the page, if not even look them up. They are an integral part of the narrative, not just references to make it legal, so to speak. I also engage in the realms of science such as time, and its beginning, to the Universe and to Life and their beginning, to how we got here and where we're going. Having gone on ahead, where to have our loved ones gone and why? I relate these crucial topics and questions back to the Bible.

We know there are hundreds of billions of galaxies in the universe, each containing hundreds of billions of stars extending over millions if not billions of light years. By contrast there are nearly ten million times more atoms in our bodies than there are stars in the universe, or grains of sand on all the beaches of the world. Even our brains are made of over 80 billion neurons, not just atoms, forming nearly one trillion connections - all grown from a single cell.

The following crucial questions present themselves at some point in most people's minds, more especially in challenging times. We will explore them together as we read on.

Who are we to say, "*a God who can create something so vast and complex **CAN** know us, love us and share His thoughts with us?*' (Amos 4:13) Why would he?

Or equally,

*"Who are we to say a God who can create something so vast as the universe and complex can **NOT** know us, love us and even share His thoughts with us? (Amos 4:13) Why wouldn't he, when 'God so loved the world he gave his one and only Son, that whoever believes in him shall not perish but have eternal life.'* (John 3:16)

These are not rhetorical questions but we should wholly consider when we explore them further.

and

What are these thoughts he has about us? "*I know the thoughts that I think toward you, says the LORD, thoughts of peace, not to harm you but to give you a future and hope.*" (Jeremiah 29:11 NIV/NKJV)

In a sense I am like one famished person telling another where to find food. I will refer quite frequently to "*the God of all comfort, who comforts us in all our troubles.*" (2 Corinthians 1:4)

Hopefully this book will comfort those feeling left behind with the comfort I myself received. So, in those times of grief, after everyone else has gone home, this may help.

Chapter 1

Science as Reality

All the matter we can see in the universe makes up a trivial 5% of everything. The rest is hidden. This could be the biggest puzzle that science has ever faced.
Brian Clegg - Dark Matter and Dark Energy[2]

There are these two young fish swimming along and they happen to meet an older fish swimming the other way, who nods at them and says, *"Morning, boys. How's the water?"* The two young fish swim on for a bit, then eventually one of them looks over at the other and goes, *"What the hell is water?"*[3] Now, I'm not the wise old fish, but the immediate point of the fish story is merely the most obvious, ubiquitous and important realities are the ones which are the hardest to see and talk about.

Our commonly accepted view of science is it should explain our reality, so we can understand what's going on all around us. We look to scientists to give us the security of a comprehensive understanding of life in the real world. However, in reality this is simply not what science is actually there to do, even in its own terms. The purpose of science is to predict the behaviour of physical objects such as stars, and galaxies, atoms and molecules and explain their origins, if it can. It does this pretty well. Science tells us in excruciating detail what matter does - but not at all what it is for.

Science in no way explains what we, as human manifestations of physical realities, are for. Ours, and the lives all around us - what do they all mean? We know intuitively there must be more to electrons than what they do. Science explains what they do, and their make up, but exactly where they came from is still a total mystery - one mathematically defined as being impossible to solve. Scientists have said quite correctly for thousands of years, *'something cannot come from nothing'*. Yet, we are here from we know not where. God is not nothing, but Science obviously has a hard time with attributing origins to God.

It's plainly unscientific to try to define laws of existence before anything existed, before nowhere became somewhere - here.

In the public mind, Science is on its way to giving us a complete account of the nature of space, time, and matter. General Relativity,[4] the best theory of the 'very big', the universe is totally inconsistent with Quantum Mechanics, the best theory of the 'very small' - of which everything is made.

We will find this word 'quantum' bandied about quite a bit as we go. What does it mean? Take a simple example of money. Money is measured in small quantities or quanta. A quantum of money is a penny or a cent depending on the currency, a clearly defined object or entity. It's the least amount of that currency that can exist. It cannot exist in any smaller quantities. Cut a penny in half and it no longer exists as money. We can come across £1 coins or $1 bills of course. They represent 100 pence, 100 cents or 100 quanta of money, but they are 'made of' pennies and cents.

Likewise light, energy and even time occur in quanta, or minimum quantities that can be no less without ceasing to exist as such. Quanta of light are known as 'photons' of a specific frequency. Electrons are the smallest quanta of electrical energy. Even time does not exist in the continuous flow it appears to be. Time exists in quantum bits of no less than 10^{-43} seconds, as defined or discovered by Max Planck in about 1900, known as Planck time, for which he was awarded a Nobel Prize in Physics in 1918. We will visit this quantum of time as being the closest we can get in time to the Big Bang itself. Before this Planck time of 10^{-43} seconds, within the 'Planck epoch" there is - no time. It does not exist, just as no money exists in value of less than a penny or a cent.

Nevertheless, the basic assumption is this, one day these challenges will be overcome. Physicists are working hard at the Grand Unified Theory (GUT) of everything, a complete story of the fundamental nature of the universe. However, in the words of Stephen Hawking, *"Even if there is ever a unified theory of everything, this will still be just a set of rules and equations - as in their origin still unknown, discovered by science not created."*[5]

But why do I even mention any of this in circumstances of bereavement?

Science Versus Faith

My purpose is to uncover the astonishing alignment between our rational, science-based view of how we and the universe got here (a place our minds might wander to at such times) with our faith and Bible-based views of it. These subjects become more relevant as one approaches a time of no longer being here and of joining those who left us behind.

We know there are thousands of billions of galaxies in the universe, not just thousands and billions but thousands of billions of galaxies. There are billions of billions of stars. So, where is God in all this? Why should he care about you and me? Why especially should he care about me?

Science tells us all the gigantic unimaginable mass and energy out there in the universe are merely 5% of what really exists. We have next to no knowledge of the other 95%. Apart from its interaction with gravity we wouldn't even know it's there. It does not interact with light or any other kind of electromagnetic radiation. It's invisible to us, by any means. It is therefore referred to by astrophysicists as 'dark energy and dark matter.'[6]

Discounting their flippant Sci-fi sound, 'dark energy and dark matter' are very real. There are many theories about what they actually are. The consensus has grown amongst the scientific community over the past 100 years that 'dark energy and dark matter' do exist. Might we just be missing something here in our understanding of God in heaven, or rather the *heavens* being in God? What do astronomers and astrophysicists tell us?

'Dark matter' and 'dark energy' do admittedly have a ring about them of basically absurd science fiction, the stuff of conspiracy theories. But they are very much science fact. First discovered about 100 years ago, they're mainstream science now, the stuff of Nobel Prizes past and future. They're even gaining recognition in the popular press, as the recently deployed Euclid Telescope peers deeper than ever before into the cosmos.

It's taken humans just over four billion years after life on earth began for us to become the ultimate, the most advanced life-form on the planet. We are described by zoologists as *'naked primates by descent, with extensive carnivore modifications.'*[7] When I consider this I'm bound to ask,

CHAPTER 1 - SCIENCE AS A REALITY

"What's going on here? Who or what then are we? Where exactly are we? How did we get here anyway? Where are we going?" Deep questions indeed, but there are times in our lives when they come to matter.

Our bodies, minds and even our thoughts and everything else around us are made from just 92 naturally occurring elements like Oxygen, Hydrogen, Iron and Carbon etc. Each element is uniformly identical everywhere it's found, at least in the visible universe. These 'elements' are the *"building-blocks of everything."* They were created in the stellar crucibles of nuclear fusion, fuelled by the original Hydrogen of imploding stars similar to, but very much bigger and older, than our own sun.

At the very moment when these giants, thousands of times bigger than our sun and millions of light-years from here, ran out of their Hydrogen fuel they imploded as supernovae at half the speed of light. Heavier elements like Iron (which constantly carries Oxygen to every cell of our bodies) Copper and Gold were created in the resulting billions of degrees of heat. Such was the power of the implosion these elements were immediately ejected from such cataclysms at 99% the speed of light. The remaining neutron stars, consisting mostly of just neutrons as the name implies, lacking the repulsive force of the charged protons to keep them apart, collapsed into incredibly dense matter. If you had a piece of a neutron star about the size of a grape it would weigh 100 million tonnes.[8]

Billions of years later atoms of all these elements created in this way basically found their way here - where we are now, making our physical bodies who we are now. They actually made it here from nowhere. Many of them are now present here in you and I. They're what we're made of. When I consider this, I'm bound to ask again, *"What in the world is going on? Are we maybe missing something between the realms of science and faith, mysteries which are apparently, but not actually, mutually exclusive?"*

By way of answers to such questions many of us have had less than satisfactory experiences of religion, of church and of church leaders not meeting expectations.

So, I will test the logical connections, such as there may be, between the awesome vastness of creation as described to us here by science, and our personal experience or perception of - church, religion and the Bible. I will look to contrast this sometimes disappointing, faith-threatening experience with the very existence of One who created all that we are, all we can ever see and the twenty times more of the rest of the universe we cannot see or know. There will be plenty of references to both.

So, these atoms of all the elements created in the stars or ejected from supernova at near light speed billions of years ago, millions of light years away are not just the tiny gritty microscopic particles we generally imagine them to be. Even this stuff - of which we are made - is simply ripples or eddies in three distinct quantum energy fields. They are the 'Electron Field' the 'Quark Field' and the 'Higgs Field' which came somehow to permeate the entire universe. [9]

Before going any further, it might be worth explaining, a 'light year' is a measure of distance expressed in terms of time. A 'light year' is the distance light travels in a vacuum in one year, which is 5.88 trillion miles or 5.88 thousand billion miles - in one year. That's 186,000 miles in just one second - always, everywhere. It will be helpful to give some idea of the vast size of the visible universe and therefore of the Kingdom of God - if there is one of course.

As it happens, I don't remember a time when I didn't know this number, as my father reminded me (or maybe tested me) many times. He was as passionate about such things and must have conveyed that to me too. Since then I've learned light never ever slows down or runs out of energy however far and however long it has travelled - for billions of years, 13.8 billion to be precise since the 'Big Bang'.

Under many circumstances, light behaves like a wave traversing the water of the sea, or like a Mexican wave traversing the members of a football crowd. Each one stands up in turn and then sits down as the *wave* moves on. But the wave is not the sea itself neither is it the crowd. The sea and the crowd transmit the wave. However, light behaves like a stream of tiny quanta of energy, individual particles, or 'photons' - manifesting themselves as a wave. The wave does not move on. Light IS the wave. In fact, light behaves as both waves and particles - at the same time. However, these are but ways we choose to try to understand things about light. Light is neither a wave nor particles alone, nor even a just a disturbance in a quantum field - it's light.

I'll try to gently introduce Einstein here without having the eyes-glaze-over effect or having to lie down in a dark room afterwards. He certainly came up with some weird stuff about Relativity and all that. As he explained himself, *"I don't really understand something myself unless I can explain it to my grandmother."* Well, I'm no Einstein, of course and you're not his grandmother, but let's do our best.

As we shall see later, in accordance with his 'theory of special relativity'[10] the passage of time experienced by a moving object slows down the faster the object travels. This has to be recognised in the programming of GPS satellites to take one example of many. Time actually comes to a stand-still altogether for an object travelling at light speed. Time stops. There is none. Light is therefore outside and beyond time, which is one characteristic of God, who is *"from everlasting to everlasting."* (Psalm 90:2) Expressed with greater simplicity, *"God IS Light."* (1 John 1:5)

Proxima Centauri, the nearest star to earth, is over 4 (four) light years away. If it exploded or disappeared, we wouldn't know until four years later. By comparison our sun is only eight light minutes away. But even so, if it suddenly exploded, we wouldn't know until eight minutes later. Before, and especially afterwards, we wouldn't be counting of course. On the same scale the moon is about 1.28 light seconds away, to give some idea of the scale of things we're talking about here.

Almost everything we see in the night sky happened millions if not billions of years ago. Telescopes are actually time machines peering at what once was - in the distant past.

Our very own Milky Way Galaxy is 150,000 light years across. Andromeda, our nearest galaxy is approximately 2.5 million light years away. Again, whatever we see apparently happening there now actually happened 2,500,000,000 years ago. And Andromeda is our nearest neighbour.

To put these wild numbers into a travel context. Driving at 70 mph it would take about 15 days to drive around the equator, five months to drive to the moon, 63 years to drive to Mars at its closest point and 4,400 years to drive on to Neptune.

The solar system alone is so huge. In 1977, NASA sent the Voyager 1 probe into space at 38,000 mph. It sounds fast, but it's only 0.0000037 times the speed of light. It didn't leave the solar system until 2012. That's going 38,000 mph for 35 years just to get out of the solar system. Even light which can go around the planet seven times in one second is slow when compared to the sheer size of the solar system. It takes eight hours for light to go from the sun to the edge. Our closest neighbouring star system Alpha Centauri is 4.3 light years away. To reach this our nearest neighbour in our own galaxy at this rate would take about 100,000 - years.[11]

To make another comparison, let's say our solar system was shrunk to the size of a penny coin (about one half inch across). The next star system would be 100 yards away. Our own Milky Way galaxy would be about the size of North America. There would be another 300 billion penny coins one inch across which represent other stars systems and their planets. Imagine for a moment penny coins are spread out all over North America approximately every 100 yards in all directions, we are just one of these 300 billion penny coins floating around in this massive sea of stars and planets beyond our galaxy. There are at least one trillion more galaxies like that.

Staggering as these facts are, most of us do have a deep sense of the vastness of the universe and our apparent insignificance in it. People of faith must also seek at some time to reconcile this science-based knowledge of the universe with Bible-based knowledge of God. But the two don't always fit well together. This is alright most of the time. Our brains are designed to hold such '*cognitive dissonance*' in check. But when the chips are down and life gets tough this dissonance can become quite distracting - as I know from experience. This book looks to reconcile the boundaries between Science and the Bible in ways that supplement and nourish subjective faith in, and knowledge of, God.

From our perspective as outside observers on Earth, we measure the time it takes for light to reach us from distant parts of the universe. However, it's important to note, time for these photons, as we understand and experience it does not exist. As we shall see in more detail later, a traveller's local timeframe becomes slower and slower as their speed increases - until at light speed it slows to a standstill. Time just stops altogether. Time no longer exists. We perceive photons as travelling towards or past us in time at the speed of light. But each individual photon itself exists outside of time because they move at the speed of light. Since they are timeless it's important to recognise they are in fact present everywhere, all the time along the path of light we eventually get to see when it finally arrives in our own timeframe.[12]

Now take this idea to its logical conclusion. Consider again, *"God IS light"* (I John 1:5) and existed before the Big Bang *"from everlasting to everlasting."* (Psalm 90:2) Arguably, all the mass in the universe was created by God from Himself. This is expressed mathematically in Einstein's famous $E=mc^2$. E is the power or energy which God invested in creating the entire visible and invisible universe, which is its mass multiplied by the humungous speed of light multiplied by itself. Is God too big to be true? The amount of the energy needed to create the universe is unimaginably huge, but so is the universe itself - yet it's there, it exists.

Just because the numbers are beyond big, beyond comprehension - so too is the universe even as we know it. And yet - we do know it. We see the stars and galaxies. They do not see us. We will come back to this.

> *"What (then) is mankind that you are mindful of them, a son of man that you care for him?" (Hebrews 2:6)* Plus, *"What a piece of work is man; how noble in reason, how infinite in faculty? In form and moving, how express and admirable? In action how like an Angel? In apprehension how like a god; the beauty of the world; the paragon of animals......this quintessence of dust?" (Hamlet Act 2 Scene II)*

As Hamlet surmises, this knowledge makes our brains hurt. It surpasses our understanding and is beyond our grasp, as beautifully expressed in the words, *"Such knowledge is too wonderful for me, too lofty for me to attain."* (Psalm 139:6)

Some say God is in heaven. I would suggest heaven is in God. He's that - big, *"He stoops down to look on the heavens and the earth."* (Psalm 113:6)

CHAPTER 1 - SCIENCE AS A REALITY

Regardless of how many billions of galaxies there are in the visible universe and however many billions **of** billions of stars, there must be countless planets on which there might be life.

Since there must be life on other planets, the argument goes, some of it must have advanced way beyond how we abuse life here on earth. So, it's kind of assumed there must be hope for us too. One day we too may get to be as advanced as they are. We've probably even heard about populating some of these life-supporting planets which might even be a solution to over-population of our own planet. Hmm.

Located at such great distances, travelling even at the speed of light, it seems more likely these planets are beyond the reach of the lifespan of humankind itself. Seeing them as they were millions of years, if not billions of years ago, how likely is it they're still even there now, not to mention when we might ever get there? One has to ask, *"What evidence is there to the contrary?"* We are effectively home alone in the universe with its Creator. The evidence is all around and even within us in the timeless particles of which we are made. More later on this too.

Despite the context it is still said, and is often implicitly assumed, *"Man is the measure of all things,"* 2,400 years after Protagoras first said it. Can it really mean because we know all this we know we really are superior to what we know? Or can there really be, can there really not be, One superior to ourselves in knowledge and in power?

Most people would tend intuitively to this view at one time or another. It has even been said, *"If God did not exist we would need to invent him or her."* However, looking up at the night sky there seems less need for invention and more to know who put it there - if you see what I mean? It has also been said, *"We tend to want to create a god corresponding to our own nature, rather than one to whom we are accountable."*

And there's a core dilemma.

Each one of us is totally special and unique, of course. Mine and everyone else's physical lifetimes are but tiny flickers in this context of billions of years of the evolution of life. And yet, we never existed before as our individual selves, nor will we ever exist again anywhere else in all eternity. Nevertheless, might we be tempted on some days to say about our lives, *"I don't always like it down here. Is this all there is? Is this it then?"*

Shakespeare has Macbeth describe life as,

> *"A walking shadow, a poor player that struts and frets his hour upon the stage and then is heard no more; a tale told by an idiot, full of sound and fury signifying nothing."* (Act 5 Scene 5)

No. God so loves the whole world he came here himself as Jesus, for us to have "life and life to the full," to die in our place - so we don't have to. (John 10:10, 3:16) The motive was love, the purpose was eternal life, and the means was the cross.

There's so very much more to what is loosely referred to as heaven, eternity and eternal life. In this unique earthly sliver of time are we in some way being uniquely prepared for something more? Might there actually be a role or a destiny for us somewhere else than in this mere 5% of what science can know and tell us, vast even as it is?

Might we put aside for a little while our own more or less painful experiences of religion and church. In the face of this awesome scientific knowledge might we be reminded....? *"In the beginning God.....created...... the heavens and the earth"* and every-thing in them. (Genesis 1) In less than 1,000 words the creation is designed to be read by all people for all time, so it does tend to generalise. It's not meant to be a shorthand version of the 'Big Bang' of what came before it, how life and evolution came about. The real point is God was there already, He said *"and it was so."* He made it all and it was all good. After creating Eve, He looked around and saw it was all *"very good."* (Genesis 1:31)

Genesis 2 goes into more detail from a ground level view. Adam was created in a wild place before being placed in, *"a garden in the East in Eden."* (Genesis 2:8) The apostle John opens the New Testament by putting the big picture into high relief. Speaking of Jesus he declares His identity with God, as God, characterising Jesus as *'The Word'* by which God spoke everything into existence. *"In the beginning was the Word, and the Word was with God, and the Word was God. He was with God in the beginning. Through Him all things were made; without Him nothing was made that has been made."* (John 1:1-4) There I would say is the answer to the question, "What came before the Big Bang?" almost by definition, since something (material) cannot come from nothing (material) but "God is Spirit." (John 4:24).

Then God Himself, the very Creator, Jesus the Word, *"became flesh and made his dwelling among us."* As 100% man Jesus became "just like one of us" in the words of the Joan Osborne song *"What if God was one of us?"*[10] And yet he remained 100% God.

Here we all are then, living in this tiny flicker of time, time which God also created from out of his own timeless existence. We cannot know all of Him of course, but He does know all of us - individually. We can have little trust in our own ability to know, but we can have every trust in His ability to make Himself known. Yet this must be done in ways which do not overpower our senses but leave us with sufficient mystery for us to freely choose. God above anyone else, who IS love, knows this about love. But, as we know, however powerful, charismatic, rich, or famous a person may be they cannot force someone to really love them. As Morgan Freeman (as

God) exclaims to Bruce in the film *Bruce Almighty* as only he can, *"If you find a way just let me know."*

So, it's been scientifically proven, time simply did not exist before the Biblical fact of this event of creation, also known as the 'Big Bang'. *"Before there was nothing. Then there was everything."*

I will have more to say about the 'Big Bang' origin of the universe from what astrophysicists describe as a *"singularity"* of zero size and vast energy, just as *"the things which are seen were not made of things which are visible."* (Hebrews 11:3 NKJV)

In this context I will also have more to say about faith, for it is, *"By faith we understand that the worlds were framed by the word of God."* (Hebrews 11:3 NKJV) *"So we fix our eyes not on what is seen, but on what is unseen, since what is seen is temporary, but what is unseen is eternal."* (2 Corinthians 4:18)

Before time came into existence there was no time before then. When I consider this, I'm bound to ask again, *"What on earth is going on here?"*

Time is a difficult subject to discuss objectively when we all live within it. Like the fish in the water. *"What, then, is time?"* To quote St Augustine, *"Provided no one asks me, I know. If I want to explain it to an inquirer, I do not know."*[13] It's easy to sympathise with him.

When did the Big Bang happen anyway?

Just immediately before the *'then'* when the 'Big Bang' happened there was no time to mark the moment when it did. In the eternity which existed before *'then'* God was there, *"From everlasting to everlasting, the Alpha and the Omega, who is, and who was, and who is to come, the Almighty."* (Psalm 90:2, Revelation 1:8) There was no past, no future. Then, Big Bang! It happened, and time began.

CHAPTER 1 - SCIENCE AS A REALITY

It is generally thought time began with the Bang of the universe *"banging"* out into space. But rather it was the very *"banging"* of space itself - as space and time, in time. It was not an explosion in space. It was the explosion OF space itself. Yes.

Before the 'Big Bang' the 'standard theory' assumes there was no space: not even empty space, just nothing. The Einsteinian picture of the universe merges space and time into a single entity: no space alone, no time alone. Common experience tells us it's impossible for something to happen in and around us independently of time. Space and time are therefore referred to as one single entity: 'Spacetime'.

It's easy to think of time as an added *fourth dimension* as just linking up time together with space to make this unified concept of 'spacetime' - just a bit of terminology. But Einstein showed time as having a fundamental, unbreakable link with the space around us as we know it - which kind of makes sense really? Our common experience tells us space simply does not exist without time, and vice versa.

It must be said, there is as yet no particularly good observational evidence to prove 'spacetime' and everything really did begin with the 'Big Bang'. It hasn't been seen under verifiable *laboratory* conditions and probably never will be. However, it is generally held to be the best we have for the foreseeable future.

There are nevertheless quite a number of alternative theories of the origin of the universe which struggle to answer the question, *"What came before the 'Big Bang?'"* Without going into detail 'String Theory' is relatively popular amongst scientists who study such things. It's a purely mathematical construct with little if any observational potential from a 4-dimensional universe since it requires ten pre-existing dimensions. 'M Theory' takes this further by adding an 11th dimension. To explain the *fine-tuned universe* which we'll come to, an infinite number of infinite universes has been suggested in which we all might exist simultaneously, or an endless series of constantly expanding universes eventually reverting to a singularity followed by a new 'Big Bang' universe and an infinite number of such cycles. See Prof Roger Penrose who collaborated very closely with the famous Richard Dawkins in popularising the original statement of the 'Big Bang' singularity in his 1983, *Brief History of Time*.

Either way they still don't address the main question, *"Where did they all come from?"*

More recently Roger Penrose's 2010 Conformal Cyclic Cosmology proposes, *"a universe which iterates through infinite cycles of Spacetime whereby each previous iteration is identified with the 'Big Bang' singularity of the next, followed by an infinite future expansion."*[14]

Although this apparently works mathematically there can be no foundation of observation of such infinite iterations of the universe. Even so it still does nothing to explain the origin of such an infinite cycle of creation and regeneration. Psalm 90:2, *"From everlasting to everlasting, you are God"*, could well describe the alternative. It becomes a matter of faith-based choice between this and such theories.

Nevertheless, Einstein's equations supporting 'Big Bang' have all been well proven by observations of many other events they predict and we use in everyday life, e.g. GPS. As more data are gathered we will either find there is nothing to disprove this concept, or some data will one day contradict the theory and destroy it.

Meanwhile, (timewise) since 'spacetime' truly all started at the beginning, there is no need to worry about why the Big Bang happened at a particular point in time. It didn't. Time itself simply did not exist before then. No wonder it's impossible for science as we know it to go there. From out of this timelessness, *"God said......"* (Genesis 1:1) and it just happened. From then onwards there was time. Before then there was not.

I look at the awesome beauty of all the other miraculous organisms around me, bi-peds just like myself, out there walking about. Everyone arrived here as part of the same billions of years journey as I did. They are just as I am, just as Charmaine was but no longer is. So, it is for all the others we've loved, who've loved us, and yet are no longer here with us - as well as for those who are. Some have gone on ahead while we have been left behind. So, are we not bound to ask - *"What in the world are we doing here? Is this it, or is there more? And if so, what?"*

So, here am I then, asking all these awkward questions from this deep experience of loss from which seems to come ever deeper insights into the nature and the ways of our Father God.

Science And Faith

Even in our sleep, pain which cannot forget falls drop by drop upon the heart, until, in our own despair, against our will, comes wisdom through the awful grace of God. (Aeschylus about 450 BC)

This quest of mine, of exploration of the alignment between science and faith, has happened not despite my circumstances. It's happened through my circumstances, as expressed so well by those prophetic words written 2,500 years ago.

Readers have, or will have, a loved one who, in going on ahead, left them feeling they're the ones left behind. Hopefully my story reflects some of this '*wisdom*' revealed through God's Goodness, through His "*awe-full Grace.*" Therefore, at some point in time, I believe this kind of experience will relate to just about everyone on the planet regardless of the level or type of their faith or level of their unbelief.

Admittedly, I'm writing from a Jesus follower, Christian viewpoint, the one I personally know best. However, there are remarkable similarities between the basic principles of most bodies of religion throughout the history of humankind. This should hardly be surprising since human groups over millennia have interpreted, each for themselves, their individual perceptions of God. The Judeo-Christian-Islamic or Abrahamic religions claim there to be one God. Hinduism and others claim there to be many. Buddhism addresses the question of God's existence as being irrelevant, the wrong question. Either way they all take a position on God of one kind or another. Science is the exception, preferring not to address the question at all, excluding it from its remit but providing instead a multitude of alternative quite speculative theories.

Every proton, neutron and electron in existence was made to be identical to every other. Not one of them is faulty or subject to random mutation. While a neutron will last for only 15 minutes outside an atom, an atom is made to last a calculated 1×10^{33} years (1 followed by 33 zeros). Compare this with the age of the universe at 13.8×10^9 (with 9 zeros).

Every creation story has a Creator in it. They appear quite differently in different cultures over the millennia. Does it not make sense, if there really is one Creator God, for the same message to vary in unique ways as it's received in each by a kind of *Chinese Whispers* process? Perhaps it's sufficient to say my personal choice, without discounting anyone else's, is for the one about whom it's said, *"God IS love."* (1 John 4:8) Out of love, God gives us perfect freedom not to love him back, to choose not to be with him. He is not just a divine predator. This freedom of choice mostly defines humanity. Jesus died to defend it as being the only way. He, too, had a choice to make in the Garden of Gethsemane - to stay and be crucified or escape into the adjoining Judean wilderness.

Despite its *'pheromone'* chemistry, Love does not conform to the scientific method.[15] It cannot be measured or consistently, reliably reproduced nor even understood but is described here in terms we can all recognise.

> *"Love is large and incredibly patient. Love is gentle and consistently kind to all. It refuses to be jealous when blessing comes to someone else. Love does not brag about one's achievements nor inflate its own importance. Love does not traffic in shame and disrespect, nor selfishly seek its own honour. Love is not easily irritated or quick to take offense. Love joyfully celebrates honesty and finds no delight in what is wrong. Love is a safe place of shelter, for it never stops believing the best for others. Love never takes failure as defeat, for it never gives up."*
> (1 Corinthians 13:4-7 TPT)

To take an example, the will of this God of Love is this, in the same single sentence He, *"forgives all our sins and heals all our diseases."* (Psalm 103:3) Jesus modelled forgiveness and healing for us when he was here. It has been said, *"God is both supernaturally natural and naturally supernatural."* He chooses how and also when to heal all our diseases, including through the medical profession of course. Healing, even of what we consider to be critical disease, may happen with or without medical intervention. It may only happen eventually by Jesus taking us home to be with him to finish the job. Either way, Charmaine and I never did have a Plan B. Plan A was never to be more than a heartbeat away from her being totally healed from cancer, right here, right now.

I'd seen and participated in healing prayer with others many times before. I prayed just about every day for nearly eight years for Charmaine to be healed - along with the daily blood thinner injections I gave her to counter the effects of chemotherapy. Some might say this was h*edging our bets* to which I would reply, with God, *all bets are off*, on a *racing certainty*. It's just a matter of when and how. I have heard it put another way, *"Belief is that God can heal. Faith is that he will."*

Towards the end of Charmaine's time here she did see herself as part of *"the bigger picture"* as she put it. She began to mind less whether she went to be with Jesus or remained here, provided God received *'the greater glory'*. This might sound like merely a kind of stoical acceptance of the inevitable. We saw her passing as in no way inevitable. She simply had such trust in God's willingness and ability to make things work out for the best in the end. He *"works all things for the good"* (Romans 8:28) as I'll explain later. And if we haven't yet seen the good then it's not yet the end. Yes, I have a few of these little *bon mots* of faith I roll out from time to help keep me going.

Sometimes, apparently for worse, but always for the better, God does not do as we expect, or at least not in ways we expect. He is God after all. I believe if things don't work out how we hope it's because in his Wisdom he has something even better in store for us. It may have to be later, when he knows we're better prepared to receive the *better* we long for without it being harmful to us when we get it. Sometimes it's a *better* not just meant for us as individuals but for others of his beloved - i.e. *the world* he so loved. Which is why I'm writing this just now.

But I do know from experience I want to share, God is drawing you ever closer to Himself. It's what he does. This was so for the first 50 years of my life without my even knowing it. Jesus didn't do all he did, in just the way he did it, for it ever to be otherwise. Everyone got died for!

I will continue to recount how science explains some of the mysteries God has not hidden from us but has hidden for us to find. Faith is totally personal and subjective.

As Denzel Washington's character in the film, *"The Book of Eli"* puts it, *"Faith is knowing something when you don't know something."* It cannot therefore be proved or disproved. However, being conducted through a *danger zone* of faith exploration I have come to see in recent years, objective scientific fact confirms, informs and enlightens biblical truth, and vice versa. Just by the way, if God does not exist, what does it matter anyway? If God does exist, how could it be otherwise than to matter - a lot?

Now, my aim is not just to reveal common ground between cosmologists, quantum physicists and theologians so they can all get along without throwing things at each other - although it might be nice. Nor is it simply to explore the depths of a 'reductio ab absurdum' so it becomes absurd to believe God does not exist. Then all the above would likely throw things at me instead. Not so nice.

Rather it's to delight in the awe of a God who IS love. (1 John 4:8) from which we cannot be separated whatever happens:

> *"For I am convinced that neither death nor life, neither angels nor demons, neither the present nor the future, nor any powers, neither height nor depth, nor anything else in all creation, will be able to separate us from the love of God that is in Christ Jesus our Lord."*
> *(Romans 8:38-39)*

But He is also Holy. He is the Creator of all things; from the vastness of the cosmos we see in the night sky to the quantum particles of the cosmos within us we don't see but of which we are all made. As we delight afresh at the reality of who God is and delight in him, *"He will give you the desires of your heart."* (Psalm 37:4)

Again, *"Faith is knowing something when you don't know something."* Faith needs no justification from science. Nevertheless, we draw strength and comfort from the Word,

> *"Now faith is the substance of things hoped for, the evidence of things not seen.... By faith we understand that the worlds were framed by the word of God, so that the things which are seen were not made of things which are visible."*
> *(Hebrews 11:1,3 NKJV)*

But God knows this. When faith is challenged such consistency between faith and science, in this scientific age, is a gift of great worth.

Everything God does is extravagant but never wasteful, although many would challenge this. Which is why the universe is as vast as it is and life on earth is as rich, complex and astounding as David Attenborough so passionately describes it throughout his life's work.

Yet there are times when storms of loss and of doubt rage through our lives, when faith either grows or it diminishes and dies. Through such storms I find that Jesus is always there with us in different and often unexpected ways, uncovering more of these objective back-stops to faith."

But why is life so hard sometimes? Why and how are we to endure all the storms of life? There are times when storms of loss and of doubt rage through our lives, when faith either grows or it diminishes and dies. Through such storms we found Jesus always to be there with us in different and often unexpected ways, uncovering more of these objective back-stops to faith.

Chapter 2

Jesus and Storms

Walking on water is just the beginning.
Shepherd - Bethel Music[16]

If you think about it, Jesus had a thing about storms. He uses his presence with the disciples in their storms at sea to show us how He is there for us in the storms we face in our own lives. In different ways, in each experience they grew in knowledge and faith, as can we.

Jesus slept through most of the first storm, then actually walked on the water in the next two. He had Peter walking on the waves of the third. Nevertheless, in all three storms, Jesus was *"with them"* as He is with you.

In each of them the disciples' faith is both challenged and deepened. Jesus appears at an ever-greater remove. From being in the boat with the disciples and yet asleep in the first storm, to getting into the boat with them in the second, he all but walks by on the water in the third. What can *The Ones Left Behind* learn from this series of storm events?

THE ONES LEFT BEHIND

Jesus in the Storm

We know God does not make bad stuff happen so he can save us from it to make us love him more. It makes no sense. Jesus did not make the storms happen, but he did not warn his disciples not to go out in them. Instead He astounded the disciples into working through, "*What's going on here? Who is this man?!*" He characterised his involvement in the three different storms in three different ways for them, and for us, to search out the answers for ourselves. He wanted them (and us) to learn how to deal with all the storms of life he knew we would have to face, and help others do the same.

It's a question Jesus addresses to each of us even now. We're challenged to choose sometimes, to step out of the boat like Peter, to walk - face to face towards Jesus, to look into his eyes, instead of at the storm - on the water. Jesus reaches out to save us from sinking - or we remain fearfully in the boat. Either way and throughout He is with us, whether we acknowledge him or not. It's who He is - Immanuel - God with us. (Matthew 1:23)

But then, in the words of the song, "*Walking on water is just the beginning.*"[15] Storms pass. Eventually we must go back to walking on the land again. Jesus heads off to Jerusalem. All we can do is ask in wonder, "*Who is that man?!*" and follow him.

One of the reasons Jesus actually came here to planet earth was, "*That we may have life and life to the full.*" (John 10:10) Shouldn't life be more as we might think Jesus wants it to be for us, more sometimes than it is?

We are commonly encouraged to think, *living life to the full* is what you do in the best *selfie* moments of the best days of your life. It can be holidays, cocktails on the beach at sunset - or whatever is our idea of *paradise*. It's what we're encouraged to strive for. Living life to the full might include the ideal car, the perfect family, the top job, self-sufficient early retirement. It might even include giving back to others less fortunate than ourselves. We seem to strive towards these goals all the time. But are we really living?

Shouldn't life be less as it mostly is - full of frustration, stress, challenges, unfulfillment, marked by bereavement, sickness and loss?

Life is renowned for the storms which do occasionally interrupt *the good life*. Or does reality seem to be more like the good life occasionally interrupting the storms? An old saying goes, *"Just jump! God will catch you. And when he doesn't - that's life!"* Does this sound rather like how it really is? Is the *life to the full* which Jesus promised really *steak on your plate* or is it just *pie in the sky* as they say?

Could this *life and life to the full* actually be intended to have some storms in it? The storms of life come and go. Might the way we deal with them, with Jesus rather than without him, determine how *fully* we live in the time between and even during the storms? After all, does Jesus not say, *"In this world you will have troubles. But fear not I have overcome the world."* He told us these things, so in him we may have peace. (John 16:33) And *"everyone born of God overcomes the world."* (1 John 5:4) To help us with this Jesus promises,

"I will be with you, to the very end of the age." (Matthew 28:20) Indeed, *"Never will I leave you; never will I forsake you."* (Hebrews 13:5)

Let's see what we can learn from how Jesus draws the disciples progressively out of themselves to show them how to deal with the storms, the physical, circumstantial and emotional storms we are all subject to.

The disciples experienced these three quite similar storms on the Sea of Galilee. Despite their life-time experience of living and working there they were terrified by all three. They must have been exceptionally violent storms to so get their attention like this. Jesus made use of each of them in three progressively different ways. He needed them to discover for themselves who he really was. From being asleep in the back of the boat to calming the storm, he went on to walking by on the waves before climbing into the boat and calming the storm, to having Peter himself walk out to him through the wind and on the waves.

Jesus used these events to teach them, and for us to learn, how we too should relate to him and deal with the storms of life by coming into ever closer relationship with him.

We know Jesus did not make the storms happen. But we see how he makes use of them to lead us towards maturity and make something of these mysterious and challenging verses:

> *"Consider it pure joy, my brothers and sisters, whenever you face trials of many kinds, because you know that the testing of your faith produces perseverance. Let perseverance finish its work so that you may be mature and complete, not lacking anything." (James 1:2-4)*

Pure joy?! A hard teaching and so counter-intuitive. Only by the gift of the faith we are given can we do it. We might discover the *"pure joy"* happening in us at the time. Or we might experience it more clearly in retrospect, as we look back over the ways in which God intruded into our lives with His Goodness on an as-required basis. Faith is a gift of God. (Ephesians 2:8) We develop it through application. We use it or we lose it, as Charmaine and I might have done over the years since her diagnosis. The doctor sat on the bed with us along with two nurses for support. I believe Jesus too was there as he promised because we had no fear, supported as we were by the prayers of others.

Fortunately, we don't have to endure life's storms alone, like orphans. Jesus promises to "be with us always" to *"never forsake us."* (Hebrews 13:5) But meanwhile, as it was for the disciples, might this life of ours be a kind of bootcamp designed to prepare us on this earth for the *"greater things"* of John 14:12. Might it also be training for our citizenship in the *"new heaven and the new earth."* (Revelation 21:1) In other words to *"reign in life"* (Romans 5:17) in this life and in the one to come, which has even been referred to as *"training for reigning?"*

This leads me to explore some of the how's and the why's of when *"God works for the good of those who love him and have been called according to his purpose."* (Romans 8:28) This implicitly confirms not all things in life turn out to be good, as we all know too well, but God promises us individually to work to turn them to good. What then is God's purpose for us in all this, over our lifetimes as individuals, and to which we are called? This is not a trick question.

Reading on to the very next verse, to verse 29 of Romans 8 we discover God's purpose for us is no less than, *"to be conformed to the image of his Son."* And isn't this what we're actually here for in this grievous world, to follow Jesus, to become like Jesus and to do what Jesus did (John 14:12) - at least?

Meanwhile, as God works in verse 28 of Romans 8, the Holy Spirit himself even assures our spirit of this in verse 16. We discover Jesus himself prays for us later in verse 34. Nowhere else in the Bible are the Holy Spirit, God and Jesus present together with us at the same time and place, such is the importance of verse 29.

Part of the tragedy of being the one left behind is this. You get to know how the life of the one who went on ahead finally ended. You get to know how they died - and they don't. All this person's hopes and plans, fears, joys and trials, the happiness and sorrow of an entire lifetime - all are gone and over with right there and then. There's nothing left to follow, no more earthly future. You get to see how it all turned out

CHAPTER 2 - JESUS AND STORMS

for them; you get to know which hopes and dreams your loved one realized, and which they will not - but they don't.

In a lifelong relationship of intimacy and trust, Charmaine and I had no secrets. When you've shared everything of the greatest and least significance with someone for nearly fifty years, the circumstances and timing of this person's death are a truly terrible knowledge. As much as you want and need to share this loss with the one with whom you shared everything - the one who died - you can not. It's perfectly illogical, but it's like harbouring a most guilty secret. Not sharing the death of your loved one with your loved one is akin to the worst betrayal of their trust. It changes your life, as they say, but not all in a bad way.

Charmaine had been misdiagnosed as not having bowel cancer. It finally emerged during an emergency, unprepared operation to prevent a ruptured bowel. She narrowly escaped death from septicaemia. The cancer later metastasised to her liver and to both lungs. She went through multiple surgeries, radiology, and various kinds of chemotherapy. There were side effects and further treatments to her voice, eyes, skin and even eyelashes. Hospitalisations for associated infections followed over the years. One of the four lobes of her lungs was removed. Later, another lobe collapsed. It was re-inflated, but then it collapsed again, hiding from view the tumour within. The doctors had long before stopped giving odds or "bandying numbers" as they put it. That she was there, still alive and really quite well was already off their scale. As God allowed these storms to happen, just as he had to the disciples, we learned through them, not despite them. As they did, we learned to know the peace of His presence.

We went through many storms like these.

But through them, not despite them, we learned to know the peace of His presence and quite literally knew no fear.

THE ONES LEFT BEHIND

Chapter 3

Again a Bride

"In Him all things hold together"
(Colossians 1:17)

Between this time and her becoming a bride for the second time (which I'll explain) we were on a stormy journey of learning with Jesus. We were inwardly at peace as things appeared outwardly to be challenging. Through it all, our eyes really were on Him. He provided total reassurance. All would be well, and even now I have to say, it still is well - most of the time.

We were married and Charmaine became my bride. As a believer she would again become a bride - to Jesus. John puts it this way, *"He is the bridegroom, and the bride belongs to Him. I am the friend of the bridegroom who stands nearby and listens with great joy to the bridegroom's voice."* (John 3:29 TPT)

> *"As a young man marries the young woman he loves, so your sons will marry you. As the bridegroom finds joy in his union with his bride, so will your God take joy in his union with you!"*
>
> *(Isaiah 62:5 TPT)*

That's what it says. That's what it means. Taking it further, Paul explains the astonishing parallel. *"For this reason a man will leave his father and mother and be united to his wife, and the two will become one flesh. This is a profound mystery* (Right!) - *but I am talking about Christ and the church"* he said. (Ephesians 5:31-32)

When Jesus came back for her to be with him they left together in their *"glorious bodies"* (Phillipians 3:21) she as a bride for the second time, as I will explain further.

Walking in Faith, Living in Hope

While these and many other Bible verses smoothed our way as *"living words"*, we walked on a platform of prayer built by so many people. As we walked in faith, we lived in hope. God lavished upon us the supernatural hope described in Romans 5:5: *"Hope does not put us to shame, because God's love has been poured out into our hearts through the Holy Spirit..."* It was like that.

They were not just words on the page. For us, in our particular circumstances, they were indeed 'living words'. Romans 15:13 (NKJV) told us, *"**Now** may the God of hope fill you with all joy and peace in believing you may abound in hope by the power of the Holy Spirit"* (my emphasis). And He really did fill us with joy and peace. Hebrews 11:1 (NKJV) says, *"**Now** faith is the substance of things hoped for, the evidence of things not seen"* (my emphasis). We were nourished by the substance and convinced by the evidence. I would even put it like this, with conviction if I may, *"This stuff works."*

These living words informed our mind-sets, our worldviews, and our daily experiences. They're summed up very powerfully in Steve Backlund's statement, which we often declared out loud together: *"If you're not glistening with hope then you're believing a lie."*[17] Some may find this hard to swallow. In our circumstances, it motivated us pretty well to seek out and confound the devil's lies. *"This is all going to end badly. You'll never make it. God doesn't really care. Did God really say ...?"* We were taught how to resist the devil. By literally laughing at his lies he fled from us. (James 4:7)

He Heals All Our Diseases

> *"Praise the LORD, oh my soul, and forget not all his benefits - who forgives all your sins and heals all your diseases."*
> *(Psalm 103:3).*

If we believe God forgives our sins, then we have no option but to believe the rest of the sentence. He also heals all our diseases. The Gospel really is good news. Healing, it is said, is not the whole Gospel, but the Gospel is not whole without it.

We pursued total healing with all our might. We adored the testimonies of healing which are so plentiful nowadays. They inspired us. Young people coming to faith have a new norm we didn't previously have. But our ceiling becomes their floor. If in doubt Jesus is perfect theology. Jesus made nobody sick. He only did what He saw the Father doing. In heaven there is no sickness. Therefore, God never made anyone sick - and never does. Jesus healed everyone who came to Him. Likewise, we are called not just to pray for the sick but to *"heal the sick."* (Matthew 10:8) We don't always see it, and it doesn't always happen when we pray, but sometimes it does and His intention is clear. This we know from John 14:12 (NKJV): *"He who believes in me will do as I have done."* Yes, that's what it says. I believe it not because it's in the Bible. It's in the Bible because it's true.

Defeated at the Cross

We also knew this. While God is in charge it's we who are in control of our lives through the freedom and authority we've been given to make the choices we make. Nevertheless, He is Sovereign. There is no contest between good and evil. It's simply not a matter of light over darkness, *may the best man win*. The best Man has already won. It was an unfair fight, as I will explain - a foregone conclusion. As C. S. Lewis puts it, *"The devil is a defeated enemy just as Hitler was defeated when the Allies secured the Normandy beachheads."* Nevertheless, there was much to fight for before they reached the inevitable conclusion in Berlin. Likewise for Charmaine, though the victory had been won from the outset there was much to fight for before she went home to be with Jesus and to receive the fullness of her healing.

Just so, Satan was defeated at the cross. As the Son of God, Jesus declared Himself quite rightly to be in this world, but not *"of this world"* (John 8:23 NIV). Satan, where he had no legal jurisdiction, instigated a conspiracy to have Jesus murdered. This therefore made his attempt illegal. Satan was guilty of unlawfully executing a plan to have the Pharisees incite the Romans to take Jesus's life. He was thereby lawfully defeated, condemned, and sentenced, as I describe later in more detail.

We have a way of empowering people through our agreeing with them. The only power the devil has is through our agreement with his lies, which we can provide without our hardly knowing it. The deceiver is very deceptive. Yet, we have much to fight for before reaching the inevitable conclusion, Jesus will return. Then, Satan will be revealed as a small and powerless being. *"Was he the one who ...?"*

God's Infinite Capacity for Suffering

God is in charge. He allows nothing - but nothing - to happen in heaven or on earth without His knowledge. In the words of a song, "*He doesn't miss a thing*"[18]. The very hairs of our heads are all numbered. (Matthew 10:30) "*Not even a sparrow falls/hops to the ground apart from the Father's will/care [for the Father is sovereign and has complete knowledge].*" (Matthew 10:29-30 AMP). This is a hard teaching.

Job did nothing wrong! It was not because he was "*righteous in his own sight*" or fearful because his children might go wrong and thereby leave a "*foothold for the enemy.*" Even in God's eyes there was "*no one on earth like him.*" He was "*blameless and upright.*" (Job 1:8, 2:3) I believe God could have prevented what happened to Job by not giving his approval to Satan to afflict him. This will sound shocking to many and even blasphemous to some. But God is Sovereign, THE ALMIGHTY! He voluntarily constrained His protection of Job through what Oswald Chambers refers to as "*providential permission*"[19] - God's "*very well then: this far you may come but no farther...*" (Job 1:12, 2:6, 38:11)

In the end Job was more than restored! "*The Lord blessed the latter part of Job's life more than the former part.*" (Job 43:12-13) Eventually after praying for his "*comforters,*" "*He had fourteen thousand sheep, six thousand camels, a thousand yoke of oxen and a thousand donkeys. And he also had seven sons and three daughters.*" Not despite the hardships but most likely through the hardships.

As we saw earlier, God did not create the storms in order to scare the disciples witless and then show off by calming them so they would love Him. It makes no sense. God did not create the storm in order for Peter to walk towards Jesus despite it. God never promised we would have a nice life. He did promise in Jesus we would have life to the full and he would never leave us. Storms happen. Jesus saves and prepares us for what is to come - in both this life and then next.

We know too, again from the famous Romans 8:28 (NIV) "*In all things God works for the good of those who love him, who have been called according to his purpose.*" Greek scholars I am told have looked into the meaning of this '*all things*' and have reached the conclusion - it means **all things.**

This verse wouldn't be there were it not expected for some less good things to happen. He promises to work on those things for the good of "*those who love him, who have been called according to his purpose.*"

But again, what is the purpose to which those who love Him are called? This is not a catch question. The answer lies in the very next verse, Romans 8:29 (NIV): "*For those God foreknew he also predestined to be conformed to the likeness of his Son.*" Through this process *of working "all things according to his purpose"* we are somehow conformed to the likeness of His Son. As in, "*We all, who with unveiled faces contemplate the Lord's glory, are being transformed into his image with ever-increasing glory, which comes from the Lord, who is the Spirit.*" (2 Corinthians 3:18) Yes, it all fits together!

In His goodness, and in His love, God sees all things, knows all things, and feels all things. Therefore, I believe God has an infinite capacity for suffering - on our behalf, just as any father or mother does for each one of their children. But as the fathers and mothers of grown up children our margin for useful intervention is limited.

As we survey our own lives, the lives of those around us, and the lives we only hear about around the world and throughout history we might be forgiven for thinking '*What on earth is going on here?!*' The forms and degrees of suffering are so far beyond measure. They are beyond all our expectations of how things should be in the world. We are numbed into experiencing a miniscule measure of it, which is just as small as the carnage going on around us is big. God sees it all, all the time - as it was, as it is, as it is to come. And yet does not come with '*the fire' to stop it.* He must truly have an infinite capacity for suffering. But, as bad as things are as we see them here, just so good must things be as God already sees them.

Love as Strong as Death

This was summarized for both Charmaine and I in another *hard* saying: "*We look to God to change our circumstances, while He looks to circumstances to change our hearts.*" We discovered in this period between the cancer being discovered and it being healed, be it here on earth or in heaven, there are spiritual, life-giving nutrients to be found.

We discovered too some of the awesome mystery in the verse from Song of Songs 8:6 (NIV), which tells us God's love "*is as strong as death.*" It tells of His "*jealousy unyielding, demanding as the grave.*" Knowing just how terrible the consequences are of a self-supporting, self-reliant lifetime here of storms without Him, Jesus chose the cross from a "*love as strong as death,*" for it not to be that way.

From letters Charmaine and I exchanged in our pre-Christian early twenties I see how God revealed to us even then, only love and death are forever, albeit at the time we were thinking more of the love. Only much later did we come to see it was God's own unconditional love which He gave us each for the other.

No Plan B

On each of those *living with cancer* days we chose not to give our agreement in any way to the *enemy of our souls,* the one who actually brought these circumstances about. We refused to do so and by default empower him. Insofar as we empower the one to whom we give our agreement. Therefore, we had no plan B.

Again, Jesus is perfect theology. If in doubt, refer to his words and actions. *"Very truly I tell you, the Son can do nothing by himself; he can do only what he sees his Father doing, because whatever the Father does the Son also does."* (John 5:19) God the Father never made anybody sick. There is plenty of sickness around already. But through Jesus He demonstrated His love in healing all those who came to him.

So instead, we gave our agreement to God. In this way we empowered the One who would use these circumstances to draw us still closer to Himself and to reveal to us ever more of His true nature.

As it is in Heaven

There can be no sickness in Heaven, so it must be said, God is incapable of making us sick. God does not just express love; He is love. It's against His very nature not to love. Sickness is the work of the devil, and Jesus came to *"destroy the work of the devil"* (1 John 3:8). Indeed, Jesus showed us how to pray for it be so, *"On earth* (even now) *as it is in heaven!"* (Matthew 6:10) That's our calling, an integral part of a Christian's job description.

In the future, in the fullness of time, God's perfect Will shall ultimately be done anyway, with or without our prayers. Jesus will return. It's *baked in* already. This has been the plan since before the beginning of time. So, there'd be no point to pray for it to be *"on earth as it is in heaven"* unless it were for now, not at the end of time and for us to be a part of it. In a later chapter, I will describe in more detail how science informs theology. Heaven, where there is no sickness, invades earth and displaces cancers and any other root cause or even remnant of sickness, physical or mental. The power of God's love holds *all things together* (Colossians 1:17) even

CHAPTER 3 - AGAIN A BRIDE

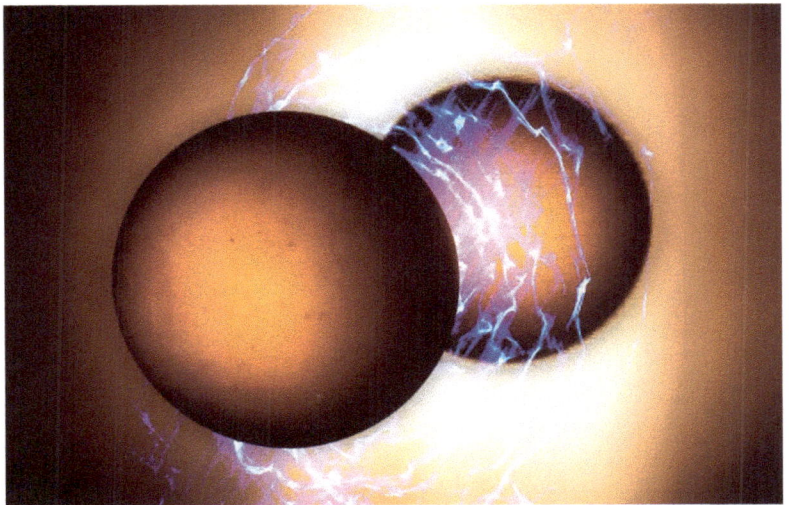

Strong Force

the '*protons*' in the '*atoms*' by what quantum physicists refer to quite simply as the *Strong Force*.[20]

Now, this does take quite some getting of the head around. I agree, we really do need power to get this. We need this kind of power, *"through being rooted and established in God's love, power, through God's spirit in our inner being.... to grasp how wide and long and high and deep is the love of God."* (Ephesians 3:18) Well, that's what it says about what holds together our bodies, our minds and all the rest of the universe. No small thing?

Towards the *"end"*, Charmaine took comfort in reaffirming she had already received the greatest healing possible. Her spirit had been redeemed, bought back, paid for at the Cross. She had *"crossed over from death to life"* (John 5:24 NIV). She knew the certainty of being adopted as a daughter of the King of kings. She would live with Him the eternal life He promised. She was content, in clear-eyed confidence, with however God would bring about this relatively lesser physical healing, either here on earth or *at home* in heaven.

If there were to be a miraculous healing on earth, she would get to see Jesus. If not, then she would still get to see Jesus. This is how we explained it to our daughter and her husband as a real win-win. Since there was no Plan B, we never ever entertained my being *"left behind."* Charmaine desperately wanted to be totally healed and to live. She wanted her grandchildren to take her to dinner in twenty years' time. Looking at them nowadays I can see how it could have been.

But whichever way it was to turn out, she wanted God to have the greater glory. I am here, the one left behind, to do what I can to *"broker"* at least as much of this greater glory as I can be trusted with, to see and witness at least some of it.

Charmaine never heard the words of this song, but she would have adopted it wholeheartedly. It speaks so much of her. *"Nothing's going to take your praise out of my mouth as long as I shall live. I will not die, I will live!"*[21] And so, with Jesus, she does!

No "What Ifs"

Seated as our spirits are with Jesus even now, *"in the heavenly realms"* (Ephesians 2:6 NIV) I believe it's impossible for nothing to happen in heaven when we pray. It has even been said, *"Prayers for which we don't yet see answers are merely gaining interest."* (Bill Johnson - Bethel Church, Redding, California) We contended constantly for Charmaine's healing to be made manifest on earth as it had already been accomplished in heaven - as did our family, our church family, and many other Christians of renown.

So, this greater glory was always front and centre in our thoughts. For better or worse then, and only by grace, we never ever felt we needed to discuss the big *"what ifs?"* Instead, we acknowledged we were part of a much bigger picture of which we had no real knowledge. Yet, and indeed, as far as God is concerned each one of us is the big picture.

"None left behind" takes on a whole new meaning as I will explore later.

The closest we came to a Plan B was a Plan A2, to bring forward a holiday to the South of France with my daughter where she had grown up and where we had spent the happiest years of our lives together. Some weeks earlier we decided to leave on January 8th. Although it turned out to become the date of Charmaine's passing, it was to be a precious time together. But for this to happen we needed a miracle and some *industrial strength* prayer.

On Eagles' Wings

On January 2nd and 3rd there was an annual conference at All Nations Church in Leicester. It was always attended by Paul Mainwaring from Bethel Church in Redding, California and two to three hundred people from the UK and Europe. Paul has a ministry specifically devoted to the healing of cancer. His aim is to partner with God in creating cancer-free zones. He'd even been given a special pair of boots with which he was commissioned to supernaturally crush the scourge of cancer.

No sooner had we arrived at the church on the Friday morning, and Charmaine had sat down in the entrance foyer, than she felt a stabbing pain in her right leg like a trapped nerve. Quite suddenly she couldn't stand up. It later became clear it was a slipped disc brought on through loss of weight and muscle tone. We could have seen this as a wicked attack to deter us, but God gave us the grace to persevere. As I'll explain later, he turned this incapacity to her good. We borrowed a wheelchair and joined everyone else in worship of our Father God. We knew that we knew, despite these or any other circumstances, He is good all the time. We would continue to worship Him for who He is, not just for what He can do for us. This, in the spirit realm, was terrorist violence against the powers mounting the attack.

Earlier the same morning a lady from our own church, Jubilee Leamington Spa, had received a prophetic picture while driving to the meeting. She saw in her mind's eye, in the spirit, a picture of an eagle in flight. Between its wings lay a person, prostrate on its back, partially hidden in its wings. As the eagle rose through rainbow colours toward the sun, these protective wings spread wider and wider. The person who had been lying there between the eagle's wings began to stand upright in joyful worship.

The lady who'd been given this prophetic "*word*" had no idea of its meaning, nor for whom it was intended - until she saw Charmaine. Then she '*knew in her knower*', the picture was for her - what appeared at the time to prophesy healing - in heaven, as I was later to learn. Paul prayed powerfully into this. Gently but firmly, he placed his booted foot on Charmaine's foot, symbolically crushing the cancer in her body. Again, although not always with manifestly physical effect, we well knew it's impossible for nothing to happen in heaven when we pray, seated as we already are "*with Jesus in the heavenly realms*" (Ephesians 2:6 NIV).

We continued to contend for the healing decreed in heaven to become manifest on earth as we experienced this season of the first storm and learned from Jesus how he worked to turn it to good.

Chapter 4

The First Storm

"Doesn't he care? We're all going to drown!"
Luke 8:22 - 25

For the first of their storms the disciples were together with Jesus in the boat, while he slept quite peacefully through it all. They despaired even though they were seasoned fishermen. They'd grown up on the shores of the treacherous Sea of Galilee. More like a large but shallow lake, it is subject to sudden, quite violent storms. Yet even the disciples, accustomed as they were to the intense moods of the lake, were afraid for their very lives. The waves were so high they washed over the sides of the boat. They were about to be swamped - and there was Jesus fast asleep in the back of the boat!

They didn't understand. They were angry and exasperated. *"Don't you care? We might drown?"* They eventually demanded he must do something, as they violently shook him awake. *"We're all going to drown!"* Jesus did calm the storm, of course. He made it stop. But perhaps as we go through life's storms there are times when we too cry out "Lord, don't you care.....? *What do you think you're doing there, fast asleep?!*" Nowadays it might be, "*Where is God whenever you want Him?*"

I once lived through similar circumstances on Chesapeake Bay near Washington DC, so I have some idea of how the disciples felt. We were out on a beautifully sunny, cloudless day. We set off on a mill-pond sea to pick up the boat captain's lobster pots. As we chugged along nicely into what became light wavelets the sky behind us gradually darkened. Weighty, menacing clouds began to appear, thickening ominously as a stiff wind somehow came up from nowhere. *"Don't worry"* said the captain, *"It happens like this sometimes. We'll be fine!"* as he headed further out into the rising waves. He wasn't exactly asleep, but his attention was away off on his lobster pots.

Then came the squalls of rain driven by a wind strong enough to turn white-caps on the waves - and we began to wonder. Two of our little group were business colleagues of mine, a couple from Sudan. They'd never even seen the sea before and had certainly never been out on it in a small boat. They were terrified, as the disciples were despite their sea-legs. *"Why doesn't the captain do something!?"* they complained to me ever more loudly. Others amongst us knew, even in our ignorance, we might have to turn around and go back. Then there'd be a real problem.

We began to understand why the captain ploughed stoically on. To turn around in this storm and head back we must at some point turn sideways on to the surging wind and waves before completing the turn. You will no doubt appreciate the risk of capsizing a boat while going broadside on to the waves. Yes, it gradually dawned on us all. But at first, '*Doesn't the Captain care?*' we all thought, '*We're all going to drown!*'

CHAPTER 4 - THE FIRST STORM

Back to the first storm on the Sea of Galilee, was Jesus just pretending to be asleep? Was he awake all the time with one eye squinting open waiting to see when the disciples were ready for him to teach them a lesson? Maybe he was hanging on tight himself, hoping they'd call him soon before it got too bad, and they all ended up in the water? I don't think so somehow. He fell asleep, as the Bible says in the story, and peacefully too I should think.

Eventually, they did wake Jesus to save the day. Shouldn't they have called on him sooner? Do we too wait until all else fails, having reached the end of our own ability to manage things? An '*If all else fails, read the instructions*' approach to life?

The longer the storm on Chesapeake Bay was left to grow while doing nothing but run before the wind, the more it took a grip and roiled the boat around in the waves. The longer we waited the less chance there was of turning safely to make it back to shore. High tension mixed with real fear was rising quite palpably. The moment of truth eventually arrived. The captain finally *woke up* to the despair in our pleas and decided to turn back. He didn't calm the storm in this case, of course, but at long last responded to our frantic calls. He revved the engine for full speed ahead, ready to power through the turn. We all willed the boat forward and braced to go quickly through the sideways-on part before heading back landwards into the wind.

Even after the captain momentously swung the wheel it seemed to take forever as we came broadside-on to the weather. Waves and spray broke over the windward side as we rolled more than steered into and then out of the perilous turn. Only as the waves again began lifting the bow did we know we'd at last made the turn and sighted the safety of land ahead. The captain had come through and we were gratefully navigated to safety.

I imagine next time the disciples were in a storm of some kind they'd think back on this occasion. They wouldn't wait quite so long but would call on Jesus as soon as the waves surged above a certain height or they began to get sea-sick, if ever they did? Might we too be mindful of Jesus being present but not active in our lives? We don't want to bother him, or admit we can't manage things on our own?

It has been said there's nothing to fear from the storm you can sleep through. And so it was with Charmaine and I. We never had a sleepless night throughout those eight years of living with cancer. It has also been said, *"If things don't turn out the way you expect, then God has something better in mind."* A bit like Job? This is a state of mind which can only be acquired as a gift through the Grace of the Holy Spirit. This is way beyond positive thinking. It's super-natural. So, *'If we're not glistening with hope then we must be believing a lie!'*

I reflect on those times, when the oncologist broke the news of another round of treatment, or something to counter the side-effects of the one before, or a fresh complication. For example, a Hickman Line is an external catheter inserted directly into the heart via a vein. Connected to a small external pump it delivers continuous doses of chemotherapy over an extended period of weeks. These doses are much smaller and less debilitating than the ones delivered over a few hours in hospital. At one point such a *line* was inserted through the vein in the crook of Charmaine's left arm. Later it was surgically inserted directly through the chest wall to minimise the risk of infections occurring from the potentially deadly incidences of septicaemia she'd experienced before.

Throughout all this Jesus was there with us in the boat, calming each of the storms as they blew up.

A Door of Hope

A few days before the conference and the vision of the eagle, before Paul Mainwaring prayed for a cancer-free zone, I'd been led to some verses from the Book of Hosea I hadn't noticed before. They suddenly seemed to assume a special significance. The written word in the Bible is a *'living word'*. *(1 Peter 1:23)* God has a way of communicating with us individually through ancient scripture with specific relevance for today. Although spoken over Israel more than two thousand five hundred years ago, these words captured my attention - for the now.

In Hosea 2:14-15 God says (NKJV),

> *"I will allure her, and speak comfort to her, will bring her into the wilderness and speak tenderly to her. There I will give her back her vineyards, and the Valley of Achor as a door of hope. She shall sing there as in the days of her youth, as in the day when she came up from the land of Egypt."*

To "*give her back her vineyards*" spoke to me of the return of the wealth of Charmaine's health. The '*Valley of Achor*' refers to a valley in Israel known as the Valley of Troubles, the exit from which would be a door of hope. The words, '*she will sing as in the days of her youth*' speak for themselves. They speak both for the youthful time of life and of the *youth* of the time when Jesus first becomes real in one's life. '*As in the day she came up out of Egypt*' speaking prophetically of release from the bondage of cancer and cancer treatment.

This had special relevance to Charmaine as her family had fled Egypt when Colonel Nasser closed the Suez Canal in 1956. Although without any possessions, they rejoiced to leave with their lives. It was like seeing God's very signature on these verses, specifically for her. It spoke much more to our hearts of the time when together we, "*came up from the land of Egypt."* We'd come out of a land without real hope into which we had been physically birthed (see Hosea 2:1-3 NIV) and into the Kingdom of Hope into which we were spiritually born again. We would experience again the *"joy of our salvation"* (Psalm 51:12 NIV) coming to know Jesus afresh for who He really is, the Creator of all things, the *'lover of our souls.'*

We the Betrothed

Although I didn't really pay attention at the time, in Hosea 2:16, 19-20 God goes on to say,

> *"In that day, declares the Lord, you will call me 'my husband.' You will no longer call me 'my master.' ... I will betroth you to me forever; I will betroth you in righteousness and justice, in love and compassion. I will betroth you in faithfulness, and you will know* [yada in Hebrew - as Adam knew Eve, as a husband knows his wife] *the Lord."*

These verses gained greatly in significance in the coming days and weeks. The Lord spoke tenderly to me in my turn of Jesus, being more perfectly balanced masculine than any man. He will come back for me one day as His bride - just as He came back for Charmaine a few days later.

On the second day of the conference, Paul's wife Sue asked if Charmaine would like to receive more prayer. But of course, we didn't know what she had in mind was for everyone there to pray. I felt moved to read out these verses 14 and 15 from Hosea - *"then you will sing as in the days of your youth."* Half of the two to three hundred people there were asked to worship the Lord, the other half to pray loudly and earnestly for Charmaine and I. Accompanied by the worship team, there was a glorious outpouring of praise and declarations of healing and wholeness. They had never done this before, but it was the industrial strength prayer we had come for. It was beyond all we could have asked for or imagined.

CHAPTER 4 - THE FIRST STORM

How wonderful it would have been to end the conference with the kind of total miraculous, lifesaving healing we hear about so often. We didn't really know at the time, but God had instead a greater glory even than this in mind. *"I carried you on eagles' wings and brought you to myself"* (Exodus 19:4) and so He did three days later.

Here is the painting the lady from our church painted of her prophecy for Charmaine, just as she saw it.

Eagle

In All Things God Works for the Good

God had always used the medical profession to minister very successfully to Charmaine. Quite unusually, we did not go immediately for medical treatment for the trapped nerve. Not wanting to miss any of the conference we didn't go home until late Sunday evening. The out-of-hours doctor prescribed painkillers. They had little to no effect. On the Monday, our family doctor wanted to be sure before we were due to leave for France on Thursday, January 8th (I'll expand on this significant date a little later) the pain was not being caused by cancer having spread to the spine. He sent Charmaine for an X-ray.

Sadly, the trauma of manoeuvring to take the X-ray caused more damage to a disc and brought on a crisis leaving Charmaine in too much pain to move. The doctor prescribed oral morphine or Oromorph. I am going somewhere with this turn of events. The following day a hospital bed was ordered for home use downstairs. Charmaine could not be moved until the Oromorph took effect.

The X-ray came through with a green light - no cancer in the bones. Hallelujah! But clearly, we would not be leaving for France on Thursday, January 8th. Our GP was determined to do whatever he could to ensure our special time together by controlling the back pain. We began to plan our departure for the Saturday instead.

Wednesday, January 7th, Charmaine's last full day here, was packed with activity. Saran, my daughter, came over from Leicester. Instead of her flying out to join us on Saturday 10th, we would drive together the same day to France - in a special Red Cross ambulance vehicle she and her husband, Julian, had arranged.

Meanwhile, we had always neglected drawing up a will. We were in touch with our solicitors over another matter and they suggested we do this on the day before we were due to leave, just for good measure. As it happened, Charmaine was not well enough to go to their offices, so they came over to the house instead that day for us to sign and for them to witness the wills. Some would see chance at work here, where I would later see God's perfect timing.

When asked by the solicitor in her absence the day before if we wanted to be buried or cremated, I opted for the latter, something most of us seem to do nowadays. Charmaine and I had never actually discussed it. I had thought *"cremation"*

CHAPTER 4 - THE FIRST STORM

was just some standard wording but when she saw it in writing the next day, she took exception when it came to sign. "*I'm not going up in a puff of smoke!*" she proclaimed. "*Anyway, I'm not planning on dying any time soon!*" she declared with panache, and she meant it. The clause was deleted and initialled. Charmaine was truly stout hearted.

Things moved quickly that day, with the solicitors, doctors, health care workers, and my daughter's visits. We rearranged our travel and even located the International Red Cross Ambulance Service to take us all the way there on January 10th.

Just before New Year we had arranged for portable oxygen equipment to be delivered. It didn't rely on heavy cylinders but extracted oxygen directly from the air. It was intended to prevent blood oxygen levels going sharply down with any exertion, making it difficult for her to breathe. The two remaining lobes of the lungs of four, which had not been removed or collapsed, were being challenged.

Our operating mode over these past eight years was this. Whichever of the countless complications, side effects of cancer fatigue, or emergencies arose, of whatever kind, we would pray. The Lord would get us over it, and we'd carry on. Strange as it may seem to read about them here, they were all just more of the same kinds of challenges. We carried on regardless, fighting from victory, not for victory.

The hospital bed arrived late on the Wednesday evening and the medical team to lift Charmaine safely into it arrived much later. It had been a gruelling few days accompanied by several nights of lost sleep since arriving in Leicester for the conference. Charmaine, while fully conscious, was a bit woozy because of the Oromorph. I'd bought a device on Amazon to clip on the finger. It was like they have in hospitals, to monitor the blood oxygen level and heart rate. I had noticed during the day how the oxygen level would go down and then recover but never to quite the same point as before. In retrospect it became clear, what remained of Charmaine's lungs were failing. But mysteriously, and quite miraculously, we were still kept from feeling any fear, let alone being gripped by it. We finally said goodnight in the early hours of Thursday morning, January 8th. I went upstairs to rest to keep going for Charmaine's recovery and our deferred departure to France in a couple of days' time.

Somewhere in the back of my mind I had wondered, if ever her lungs were to fail, how would she get through the physical panic of not being able to breathe? The Oromorph was brought in to deal with a strange and otherwise inexplicable *attack* at the very entrance to the church where we'd gone for a last-minute miracle. In the end, I believe it allowed a peaceful, if totally unexpected, pulmonary or cardiac arrest to occur without any distress. *"In **all** things, God works (my emphasis) for the good of those who love him and are called according to his purpose"* (Romans 8:28).

How do I know, and again what is His purpose? Well, it was not without considerable soul-searching over the following days and weeks, as you can imagine. But the position in which I found Charmaine later on the fateful morning was totally without distress. It was as if, having left the bed she'd misjudged and fallen to the floor beside it. After placing her head comfortably on her folded right arm there she was - just waiting for me to come back. I'll return to the reason for this amount of detail.

Never did we relent in our hope of miraculous healing. Charmaine's very last text to friends and family read, *"Mike's taking me away on holiday next week."* Bless her. Nor did we languish in the despair of *what if*. We did not embrace death as God's will as a kind of victory over sickness, but neither did we cling desperately to life in this world to avoid defeat. Death is not automatically defeat. Worse for some than death is a life of despair. In military terms it can be a tactical withdrawal to become the kernel of wheat which falls to the ground, designed to achieve strategic ends. As Jesus puts it, *"Very truly I tell you, unless a kernel of wheat falls to the ground and dies, it remains only a single seed. But if it dies, it produces many seeds."* (John 12:24)

Twice the devil lost! We knew no fear. Jesus was with us in the boat. He calmed the storm in our hearts. The cancer died. Charmaine lives.

Once again, I was taken to *"the purpose to which we are called"* of Romans 8:29 *"For he knew all about us before we were born and he destined us from the beginning to share the likeness of his Son."* (Romans 8:28 TPT) And so into the season of the second storm.

THE ONES LEFT BEHIND

Chapter 5

The Second Storm

He was about to pass them by.
Mark 6:48

In the second of Jesus' storm scenes, he arranged to raise the disciples' game as we'd had to raise ours. It wasn't quite as stormy as before but still the disciples were *"straining at the oars because the wind was against them."* (Mark 6:47-51. John 6:21) Instead of being close at hand, to be called on at any moment, Jesus hadn't joined them in the boat this time.

But, *"shortly before dawn, he went out to them, walking on the lake."* He was, *"about to pass them by"* unless and if they called him. Later in life, with Jesus no longer at arm's length in the back of the boat we, have to reach out to Him in order to really grow.

Despite, or because of his love for us, Jesus doesn't always come round checking to see if there's anything we need. Although always present, since *"in Him we live and move and have our being"* (Acts 17:28) He's no celestial *bell-hop*.

THE ONES LEFT BEHIND

It has been said, *"While God provides what we need, what we want we have to ask for,"* in order to grow to maturity (Bill Johnson, Bethel Church, CA).

'They thought he was a ghost.' Only then did they cry out, and even then, it was because they were terrified of the *ghost*. Otherwise, they might well have continued *"straining against the oars,"* still in their own strength.

> *"They cried out because they all saw him and were terrified. Immediately Jesus spoke to them and said, "Take courage! It is I. Don't be afraid." Then he climbed into the boat with them, and the wind died down. They were completely amazed." (Mark 6:51)*

It's as if Jesus makes use of the circumstances of the storms of life to teach us not only about his authority to calm them but also to minister to our own spiritual and emotional needs.

So, it had been with Charmaine and I. Jesus calmed the eight years of storms caused by her cancers as we called on him, in or from the boat. We learned bit by bit over the years. Jesus was not just Lord in our lives but Lord of our lives.

Even from the outset the trusted oncologist told us, *'We're both going to lose here. The cancers are such I cannot save your life. I won't bandy numbers with you, but I can only prolong your life for a while.'* But Charmaine and I had been taught through Jesus's use of circumstances, as the disciples had, to *"Take courage"* over fear, over the terror of the cancer and its possible consequences. All throughout, the only fear we had was of fear itself, so to speak. Jesus dealt with our fears as he walked to us over the water. He calmed the storms and climbed into our boat.

Chapter 6

The One Who Went Ahead

You came I knew that You would come
You sang My heart it woke up
I'm not afraid, I see Your face I am alive
You came I knew that You would come.
"Lazarus"- Bethel Music

As I said earlier, part of the tragedy of being *The One Left Behind* is this. We get to know how life ended for the one we'd spent our lives with who went on ahead. At least part of my struggle now is still loving Charmaine and not being able to tell her. It's like the betrayal of a most sacred trust. To have intimate knowledge of how the life of someone so close to you came to its end and not to tell them of it.

To know and to be known in such a way is such a precious gift, and I hardly knew I had it. Yet, there was not a single plan, ambition, or dream of the past fifty years we had not shared together. Charmaine faithfully validated each one of them for me. She somehow made them real even before the event.

Why Wasn't I There?

Now she's gone, and still I hardly know where to put myself. Rather like having the pain of a phantom amputated limb, only sometimes it feels more like half my body is missing - as I'm sure some of my readers will understand.

What's the point of anything? I am still having to learn a whole new way of doing life. It may seem melodramatic, if not trite, to say so. But, some years on I realised this, my heart is actually broken. This is not to say I'm brokenhearted - that too, I suppose - but rather my heart doesn't seem to work as it used to. Sometimes it's as if, *"I've stopped breathing but I'm still fully aware,"* as so graphically expressed by Lady Gaga in the song *A Million Reasons*.[22] Amongst the million reasons to let go she finds the one reason to stay.

But God did not allow Charmaine to go ahead and allow me to stay behind without there being a purpose for it, unique bigger picture ways to advance His Kingdom - whatever they might be.

Meanwhile, pressing urgently on my mind was the need to understand just what had happened that night. Why had Charmaine left the bed? Had she just fallen? What was she thinking? What were her very last thoughts? Was she distressed? Was she afraid? Did she know she'd have to go alone? Did she cry out for me? Did she cry out to Jesus? Why was I not there? What might have been our last words? Why did God not wake me? Why did she have to go at all? Why hadn't I prayed more? Why hadn't I fasted more? Why couldn't I fix it so it didn't happen? Why wasn't I there?

CHAPTER 6 - THE ONE WHO WENT AHEAD

At Rest in His Love

But, God is Good. A little later on, he revealed to me what happened. He explained it to me in increasing detail over subsequent weeks. I was struck by the well-known John 14:1-4 which took on perfect relevance. There Jesus encourages us,

> *"Do not let your hearts be troubled. You believe in God; believe also in me. My Father's house has many rooms; if that were not so, would I have told you that I am going there to prepare a place for you? And if I go and prepare a place for you, I will come back and take you to be with me that you also may be where I am."*

In a way I cannot explain, the Lord showed me Charmaine had left the hospital bed as she did to be with Jesus. He had come back for her just as He'd promised He would. She saw Him coming to take her home to His Father's house as His bride - and got up to meet Him. Yes.

A few weeks later, I knew for certain - Jesus had at last *"come leaping, over the mountains, bounding over the hills"* (Song of Songs 2:8) to claim His bride - just as I had myself in years gone by. He just couldn't wait any longer to take her to be with Him. The Lord took me to Martin Smith singing the *"Song of Solomon"*[23] In this context, it speaks of her for itself.

> *Do not hide me from Your presence Pull me from Your shadows, I need You. Beauty wrap Your arms around me Sing Your song of kindness I need You*

So it was I found her as it had been in Song of Songs 2:6 (TPT) *"His left arm cradles my head while his right arm holds me close. I am at rest in his love."* Just so, Charmaine's head was supported on the right side, as it were by Jesus's left arm, and it was as if she lay in the embrace of His right arm with these words still on her lips.

> *"With you I will go."*
> *All through the valleys*
> *Through the dark of night*
> *Here You come running to hold me till it's light*

Through the dark of night, Beauty had come to wrap His arms around her. If even I could see it, then she must have known. I was consoled in not being there. God is Good, and He is Kind - all the time.

In Song of Songs 2:10 (TPT), he showed me how He'd said to Charmaine,

> *"Arise my dearest, my beautiful one. Hurry my darling. Come away with me! I have come as you have asked to draw you to my heart and lead you out. For now is the time, my beautiful one."*

I *knew* they had left together arm in arm, bride and Groom, just before I got there. What remained was the kindest way Jesus could have shown me His love for her, and for myself even though He'd taken her to be with Him. There was no distress, no striving, no despair in the way she lay, just perfectly at peace. She was *'at rest in His love.'*

Let Her Go Now She's Mine

Some months later, I took the trip alone to Israel we'd planned to take together. We were baptised some years before in our church pastor's swimming pool near Nice. We were living there at the time where we had come to know Jesus as Lord.

Some years before then, my father died of cancer and my mother followed six years later. One of Charmaine's cousins is a full-time evangelist. His advice was to read the Bible every day. The good news was I didn't have to understand it. I was a good practitioner of Buddhist meditation at the time, with an emphasis on being in the present. I just accepted things as they are. I'd begun to view the Bible in a non-threatening

CHAPTER 6 - THE ONE WHO WENT AHEAD

way as just a book, like any other and something you could just read. Not unusually, I'd had pretty much a horror of it until then - as many do. So, I thought I could about manage this, especially the not understanding it part, and read the New Testament pretty *religiously* just about every day for four years - without understanding it at all.

However, I did notice this. Jesus made frequent reference to his fulfilment of scripture from his birth to death and resurrection. I was quite struck by the shear audacity of this man who would make sure to fulfil all the prophesies about the Messiah in the Old Testament right through to being crucified *"pierced for our transgressions"*. (Isaiah 53:5) Who would do such a thing as go to the full extent of dying such a death - in order for people to think he was God - and then only after he'd gone? I pondered the mystery of these verses:

> *"Who, being in very nature God, did not consider equality with God something to be used to his own advantage; rather, He made himself nothing by taking the very nature of a servant, being made in human likeness. And being found in appearance as a man, He humbled himself by becoming obedient to death— even death on a cross!"*
> *(Philippians 2:6-8)*

I puzzled over this for many months until horror of horrors I found myself agreeing with Charmaine to take me to a local Anglo-French church. God had made use of some serious financial difficulties of ours to get our attention.

I needed to know how I could be sure about Jesus. Was he really God and not just a clever, over-ambitious and rather fool-hardy imposter? The somewhat banal and rather obvious but life-changing answer came from the pastor. He quoted Jesus' words to Phillip in answer to the same question, *"Don't you know me, even after I have been among you such a long time? Anyone who has seen me has seen the Father."* (John 14:9)

And that, peeps was that! I believed the whole thing - everything. It had only taken about 40 years to get there to the last step and the beginning of the next step(s). Jesus became not only Lord in our lives but in time he became Lord of our lives - Charmaine too - at the same time, although I think she'd been discretely holding back for me to get it first. Such were the pressures of potentially losing our home through poor financial advice leading to non-payment of government taxes. But there's another story.

So, back to the months following Charmaine's going on ahead. In May of the same year, we'd planned to be rebaptised in Israel, in the Jordan. Jesus released me in a way from the depths of grief at revisiting Israel and some of our most favoured places. Charmaine and I had adventured there in such joy just years before. Standing there idly in line, waiting by the Jordan to be re-baptised I *heard* very clearly, in my mind, "*Let her go now; she's mine.*"

Had it been my own thought, it would most likely have been something nice like, '*It's all right now. You can let her go. She's with me.*' But it was said with such authority and above all, with such depth of love, compassion and understanding. It could not have been my own thought. No, it was just, "*Let her go now she's mine!*" He showed me Charmaine had become fully His.

A few weeks later someone drew my attention to a verse I'd never noticed before in Isaiah 43:1, "*I have summoned you by name; you are mine.*" Having promised in the next verse, "*When you pass through the waters, I will be with you; and when you pass through the rivers* (as I most certainly would) *they will not sweep over you.*"

This gradual revelation had been at a pace I could deal with. The Jesus Culture song *Dance With Me*[24] again evokes a glorious image of Jesus coming to claim His bride, paraphrasing Song of Songs 2:8, 11-12.

CHAPTER 6 - THE ONE WHO WENT AHEAD

Won't You dance with me, Oh, Lover of my soul, to the song of all songs? With You, I will go You are my Love You are my Fair One The winter has passed and the springtime has come

The season of singing as in Hosea 2:14-15 above had come. The *winter* of eight years of sickness had indeed passed, and for Charmaine the eternal springtime had indeed come. The Passion Translation speaks of it in Song of Songs 2:10-12 (TPT):

> *My beloved spoke and said to me,*
> *"Arise, my darling,*
> *my beautiful one, come with me.*
> *See! The winter is past;*
> *the rains are over and gone.*
> *Flowers appear on the earth;*
> *the season of singing has come,*
> *the cooing of doves*
> *is heard in our land."*

They had left together hand in hand, bride and Groom, to the sound of doves to His, to their land.

But Where Is She Now?

God knew I needed to know. Where is she now? He was teaching me. His *"love is as strong as death, his jealousy demanding as the grave."* (Song of Songs 8:6) One way He revealed it to me was like this.

In preparation for our wedding, we'd planned for Charmaine to join me in Manchester, to live temporarily at my father's house. On March 22nd 1972, I wrote to the registrar of marriages in Manchester requesting permission for our wedding to take place on May 6th, her birthday. He replied saying it just couldn't happen because Charmaine had not been resident in Manchester for long enough. I didn't discover this letter until the very same day of March 22nd 2015, a year to the day having completely forgotten about it.

In the letter, I mention how Charmaine left her home on January 8th 1972, to come to my father's house in Manchester to be my bride. As it happened Jesus returned for her to take her as His bride to His Father's house, in the Jewish tradition. It was exactly forty-three years later to the day, on January 8th 2015. Somehow, I know in my heart, where it counts for at least as much as in the mind, these dates and this letter are beyond coincidence. I take them as documentary *proof* of God's faithfulness to His word - to the very day(s)! Such a *God-incidence*.

I know Jesus came back to take Charmaine to His Father's house so she too may be where He is. True to His promise, she was twice a bride!

God has often spoken to me through numbers, catching my attention in this way long before I knew Him. In the very first letter Charmaine wrote to me the telephone number was Harden 576. Our birthdays are in May (the 5th month) on the 7th and 6th days. A few years later I was on an assignment to Swaziland. We self-separated for a year just before we were married. The number on my post-office box was 47, my birth year. Have you ever asked about such things, "*How do you do it Lord?*"

Sometimes I still wonder to myself, '*Where are you now?*' As one who was left behind would ask. Again John 14:3 affirms, "*That you also may be where I am*" - with Jesus. It's as simple as that for those who have gone ahead. Like Jesus said she is, "*where I am.*"

Further, it says very clearly in Ephesians 2:6 when we came to believe, came to faith as some say, "*God raised us up with Christ and seated us with him in the heavenly realms*" no less!" Charmaine is there now, where together we have both been ever since then "*with Jesus in heavenly realms.*" Our separation here on earth, although real, is only apparent.

Even to some Christians this may seem a bit outrageous. Yet as a scriptural fact it should be of comfort. And *"Far be it from me to not believe (what is written) even when my eyes can't see."*[25]

What then, and where are these *"heavenly realms?"* became important questions to me, as you might imagine, and as they might well be to anyone else who's apparently been left behind.

God Is Not in Heaven

Generally speaking, most of us just think of God as being in heaven. Everybody kind of knows and tends to look upwards at the appropriate time to a *better place*. But then I believe God is very much bigger even than we think. I would go so far as to say God is not in heaven; heaven is in God. *"Heaven is my throne."* (Isaiah 66:1) In a later chapter, I will explore further how scientific fact converges with Biblical fact.

The visible Universe with its billions of galaxies and billions of billions (not billions and billions) of stars is but 5% of the whole of existence. The other 95% is described as 'dark matter' and 'dark energy'. They literally co-exist with the 'normal matter' and 'normal energy' of which we are made. Could this be the *invisible* part of creation? (Colossians 1:16) We can neither see nor detect with instruments what 'dark matter' is since it manifests simply through its gravitational impact on everything else. The Euclid Telescope was launched on July 1st 2023 to explore further.[26]

All through Charmaine's last summer, she was reading the gospel of John with a prayer partner. They were struck by Jesus's references to His time not having come. Then, in the autumn, and especially her final weeks, she was impressed by John 12:23, *"The hour* (the time) *has come ..."* Jesus declared quite often over his years in ministry that, "My time has not yet come." Likewise there were many times too when there were improved test results, fewer, less severe symptoms when we were sure "the time had not yet come" so to speak. Not only that but it never would come. Although we never spoke of it, I came later to wonder myself if Charmaine hadn't begun at some point to wonder to herself, "maybe a time will come....." as one does. She identified with all the verses which appear later where Jesus said, *"Ask and it will be given to you."* It was the time to ask. So, we asked, and asked, and continued to ask - that His name be glorified as she submitted to her small part in His *big picture,* as she put it.

As far as God is concerned we are, each one of us, the *big picture.* He's like that, and while not wanting to sound glib, it has been said, "Did you not know? God has a picture of you on his refrigerator." I think there's truth there about God's character, who IS love.

Returning later to these verses in John 12, I discovered a deeper meaning. I cannot say Charmaine had picked it up, and we never discussed it. These are Jesus' words about Himself, but I think here as elsewhere, when revealed by the Holy Spirit, they can be justly applied to ourselves.

Verse 23 says, *"The hour has come that the Son of Man be glorified."* Had we not agreed? Whatever the outcome, God should have the greater glory?

Verse 24 says, *"Very truly I tell you, unless a grain of wheat falls to the ground and dies, it remains only a single seed. But if it dies, it produces many seeds."*

Verse 26 says, *"Whoever serves me must follow me, and where I am my servant also will be"* in heaven with Him.

Verse 27 says, *"Now my soul is troubled, and what shall I say? 'Father, save me from this hour?' No, it was for this reason I came to this hour. Father, glorify your name!"*

Certainly, we were *"troubled"* by the advancing symptoms of weight loss, breathing difficulty, tiredness, loss of appetite, even inability to eat sometimes and loss of mobility. But there was this powerful, overriding sense of there being a "*purpose*" for it all.

I now have stewardship of this seed, amongst others for whom it was left. It's for us to water it and for the Lord to make it grow. (1 Corinthians 3:6) I hope these pages will show how He did, and is, doing so - perhaps in your life too. Surely it was for a reason for any of us to come to this or any other hour - for our Father to glorify His name - sometimes in ways known only to himself.

I want to share the spiritual nutrients we discovered in those years between the diagnosis and the ultimate healing for which Charmaine went *home* with Jesus to complete. When Jesus seems no longer to be in the boat but lost to us - out there somewhere, as we shall see later in the Third Storm, we need to learn how to be nourished at such times.

THE ONES LEFT BEHIND

Chapter 7

And Now?

> *"The Spirit of God has made me,*
> *And the breath of the Almighty gives me life."*
> *Job 33:4*

From receiving the news in mid-November of just three months to live, it turned out to be less than two. The family doctor arranged to call by very early on that last morning of Thursday 8th just to check how things were going. Even he was shocked to find Charmaine was longer with us.

Why!?

Having dealt with the preliminaries of informing close family and the undertakers, reality began to break in hard upon me on Friday, the following day. I began then, and continue to be thankful, Charmaine herself was not *The One Left Behind*. I would in no way want her to go through what awaited me. I soon had to come to terms with referring to *I* and *me* instead of *us* and *we*. Among many kindnesses someone pointed out I am never alone. It's never just *I*. With the constant presence of the Holy Spirit, it's always *we*. I haven't really worried about making the same distinction ever since.

Having completed the funeral arrangements on the Friday, my daughter Saran and I were determined in some way to honour God's timing. We flew out to France on Saturday the following day on the very flight she had always planned to take. God's timing is the signature of His presence and His kindness. Somehow, we arrived at our destination on time - although not quite as planned. What a blessing in some ways it was since Charmaine's decline was so rapid. Saran had come by on the Wednesday. The three of us had been practicing our French, and went on to spend a unique and precious time together.

Nevertheless, I did begin to get ticked off with God. Looking out from our balcony across the glittering, azure bay of Villefranche (near Monaco, just outside Nice, where we'd lived for fifteen years) through the palm trees under a perfect blue sky, I began to ask this, *"Why could she not have lived just a few more weeks? Why could we not have shared this first and last opportunity to be here with our only daughter for the first, and if really necessary the last, time in nearly twenty years? Just look how eye-wateringly beautiful all this is! Why, Lord?"*

From Glory to Glory

I don't know whether you've experienced this yourself, but I've found God has a way of speaking so you know it's Him. His voice is always full of wisdom, full to overflowing with love, but with the gentle and inescapable authority which not only persuades but also convinces. So here it came. *"Would you ... would you, Mike, compare what you're seeing now"* - and you can see where this is going - *"with what Charmaine is seeing now?"* And that was it, my objections were toast, and it was suddenly OK in my heart. His words are accompanied by effects we cannot achieve just by positive thinking. I came

CHAPTER 7 - AND NOW?

instantly to the conclusion too, perhaps on my own, it wasn't any of God's fault we hadn't done this before.

As I *reverse engineered* the events of the past few weeks, I could see God's hand in them all. I thought of how it might have been: lingering between life and death, destroyed by bone or brain cancer; a husband caring for a wife in ways he never had to before; incarceration; agonizing daily visits to the hospice; neither of us ever knowing if this would be the last good-bye. Instead, Charmaine received the "*glorious body*" which will not see decay, the one Jesus promises we shall have (Philippians 3:21) which I will describe in more detail later on. But I rejoice in this even now. God was coaxing me, *wooing* me "*from the jaws of distress to a spacious place*" (Job 36:16) to come into agreement with the wonder, the mercy and the kindness of His ways.

The Lord allowed Charmaine to go ahead of me. However unwillingly, this allows me the freedom to do things we would not otherwise have done together. This is just a fact, and in it a burning question, beyond writing this book, *What are they going to be?* My move to British Columbia to be with our daughter, her husband and their family, our grandchildren - is full of promise.

Either None of It's True or It's All True

Meanwhile, the moment you die is a very important time in your life, so to speak. This is also true of the person who remains who spent most of theirs with you. They get to live on and reflect upon it. Did it go OK? Should it have gone differently? What might we have done differently? What might I have done differently?

When you find the person you know and love so deeply is so clearly no longer alive, when they always had been, somewhere in your mind you really begin to wonder, is all this about God really true? Did they just simply stop existing? Tell me, is there really life after death, like they say? Does eternal life in some timeless place really happen, or is it just a big *Nothing*, as the lifeless *Nada* of a body lying before you seems to be? The love of my life, lying there disguised as a dead person? I needed to know.

What more might I have done had I not become so spiritually fatigued over the past eight years, and especially the past three to four years of Charmaine's and her mother's terminal illnesses? Her mother had passed away less than nine months before.

Not that I didn't pray. I prayed just about every day, with Charmaine and on my own. The cancer had no right to her body. I commanded it to leave in the authority I'd been given in Jesus' name, and we likewise declared it had already left. But could it have been with so much more persistence, more *violence!?*

Knowing what I know now, there's no way I can blame God. But knowing exactly what I know now, I have to consider very carefully my own part. I've heard it said, *"The hardest challenge of the Christian life is to navigate between blaming God on the one hand and one's own guilt and shame on the other, without foundering on either or even both."* As Charmaine used to say, *"Either all God says in the Bible is true, or none of it's true. There is no in between."*

As I've begun to describe here, everything within me cries out, *"I believe! It's all true!"* Thankfully I find this is so not just because I want it to be. It's confirmed by so many of the blessed personal events of the times I describe here and so many others over my entire life. They impress themselves on my heart. When I reflect on the wonders of the universe, of Creation I've already alluded to, that we are here at all, how could I not - believe?

CHAPTER 7 - AND NOW?

A Death to Glorify God

So, dare I even suggest God was preparing our hearts for ways in which He would glorify His name? Although the circumstances are completely different, again perhaps there really is *"a death that will glorify God."* (John 21:19)

But this kind of death, Charmaine's kind of death, is no defeat but a victory of the childlike faith Jesus advances, to wait on God for His best. Some of the following paradigms originate from Bethel Church, which we adhered to and regularly declared. Trite as they may seem, they are empowered by prayer and God's Grace:

> *If you're not glistening with hope you're believing a lie. If things don't work out as you expected then God has something better in mind. If we don't see answers to prayer just yet it's because they're gaining interest.*

Charmine received what's known as a word of knowledge for healing from one of Bill Johnson's team members from Bethel Church while visiting Cardiff in 2008. While addressing an audience of close to 1,000 people his assistant felt impressed upon her mind that somebody there that night had what she referred to as a 'spot on the liver.' This is known as a Rhema word where the Holy Spirit reveals to someone what turns out to be true about another person. They may or may not be identified at the time, maybe by name or something they're wearing. Charmaine knew in her heart that this was a word specifically from God for her, recognising that He was "on her case" as she put it. As a result, she was the only person to go forward for prayer for this condition and drew great comfort from it that God had expressed His intention of healing her from, 'the spot on the liver.' It was exactly the same term the medics used to describe what they saw. The liver was healed through surgery. However, she did notice there was never a Rhema word for the lung cancer, the healing of which she had to go to be with Jesus.

Breathe on Me, Breath of God

We were given another prophetic word in mid-October 2014 during five days we spent at a Christian retreat in Wales called Ffald-y-Brenin.[27] Quite without thinking, Charmaine heard herself say one evening, *"Breathe on me, breath of God."* Certainly, to have more of God's breath in her lungs would have been good.

The next day one member of a prayer team there shared, *"The only thing I have for you is, 'Breathe on me, breath of God,'"* which was quite amazing really since we had not confided the experience of the previous evening with anyone. Later, we did tell a different staff member about this *coincidence*. When she got home at the weekend, she was surprised to find a CD still on her player from the week before with the song on it, *Breathe on me, breath of God*.[28] What kind of a *coincidence* is that?

The first verses of the song are quite encouraging, as they speak of healing and restoration.

> *Breathe on me breath of God Fill me with life anew*
> *That I may love What thou do'st love And do what thou*
> *would'st do*

In retrospect, the last verse speaks prophetically more of heaven than of earth.

> *So, shall I never die but live with thee*
> *the perfect life of thine eternity.*

As Jesus himself said, "Whoever lives by believing in me will never die." (John 11:24)

Some might actually see this, on the face of it, as a kind of deception, which is why it is so important to know the character of God. The Israelites only knew *"God's deeds"* what he had done. Moses *"knew His ways"* (Psalm 103:7) he was intimate with God's character. Our job too is to know

His ways, His character, to know Him, to be assured of His enduring goodness.

A time came when Moses, having fled Egypt as a murderer, was assigned to return there 40 years later to lead the Israelites to the land of promise, the promised land. But Moses, described as a friend of God, having such a relationship with God as did Abraham, would sometimes bargain with him.

So the story goes, Moses quibbled with God, *"Why me? Who am I to risk death for everyone by opposing Pharoah?"* God replied by asking Moses, *"What's that in your hand?"* *"Just a stick,"* stammered Moses. *"Fine,"* said God, *"that'll do."* (Exodus 3:11,14:15-16)

Much later on, after judicious use of this '*stick*' or '*staff*', to part the Red Sea having witnessed God's power in destroying the Egyptians and much else, Moses needed reassurance. In a fascinating story in Exodus 33:17-20, before heading out from the other side of the Red Sea to the promised land, Moses had another wobble. He asked God to show him His Glory. But God knew Moses really needed reassurance of even more than his glory. He replied instead, *"I will cause all my Goodness to pass in front of you."* (Exodus 33:19)

What I read from this is God's Glory is his Goodness, not just his glorious deeds, but the goodness of his ways, his intentions as well as his accomplishments. In those times of plagues and mass murder, Moses needed to know. Who is this God I've become so involved with, just as it is for each one of us in times of sickness and bereavement.

While at first it may sound trite, I would even put it this way and live by these rules as a lifestyle:

> *Rule #1: God is Good all the time - and He is Wise*
> *Rule #2: If it ever appears God is not Good and Wise,*
> *see Rule #1*

By this and through it all I was encouraged. At the same time, and in very practical ways, He showed me signs of the wisdom of his goodness, things he knew I'd find later, things hidden for me to find, not things hidden from me. They were to demonstrate He really had been there all along, and so He is still.

Here are some examples: Charmaine leaving her family house for my father's house on January 8th 1972 as on January 8th 2015 she left mine for her Father's house - in heaven.

When I found the letter I wrote to the registrar of marriages in Manchester on March 23rd 1972 on the same day in 2015.

When I drove with my daughter to the funeral from one part of town as the hearse carrying Charmaine was driving there from another. It passed last through the traffic lights from one direction just a moment before we passed first through the same traffic lights from the other direction as they changed. We drove right behind them for several miles, one behind the other, together with them out of town right to - the crematorium.

Coincidence, some may say. But in these pivotal circumstances one actually feels blessed even beyond the odds against such things happening. It's as if God is saying, *"That was Me. Don't even ask how I do it, just be blessed."* I will explain later on how the same God who shows us his goodness in so many ways both created and sustains the entire universe even as he blesses us. Who are we to say he can't do both? Who are we to say he can't - when he does? What else should we expect of *'the living God'* if he is not to be more than a god?

The peace we had could be seen by some as simply denial on our part - *'what's meant to be will be.'* But God allowed Charmaine to go so peacefully and painlessly, as and when He did. I have little choice but to believe He had been preparing our hearts for what was to come. I most certainly feel as though we had been prepared, both for Charmaine's part and for my

CHAPTER 7 - AND NOW?

own. Otherwise, by now, my life would be a wreck, having simply *rearranged the deckchairs on the Titanic*. Praise God it simply is not, even though at times it might only seem like it.

In the end, and as the January 3rd 2015 prayer at All Nations Church, Leicester showed, you can't pray successfully against God. His timing and His good and perfect will prevails. He is not in control of how we respond to events. He has yielded this prerogative to us. But He is ultimately in charge.

Jesus, speaking here of sparrows says, *"Not one of them will fall to the ground outside your Father's care. So don't be afraid; you are worth more than many sparrows."* (Matthew 10:29 and 31) Trite indeed if taken simply as read. But, like all scripture, when the right verse is imbued with the power of the Holy Spirit for the right person at the right time - the fear goes away.

With Jesus no longer with us in the boat, nor even anywhere in sight, we were far from shore. But I couldn't quite imagine Him saying, *"Sorry, pal. Thanks for playing. Charmaine's just outta here."* That's not the Jesus I know from the third storm and much more.

THE ONES LEFT BEHIND

Chapter 8

The Third Storm

Whatever you do, don't look down.
Anon

At the time of the third storm, the disciples' boat, *"was already a considerable distance from land, buffeted by the waves because the wind was against it."* (Matthew 14:24 28) *"Shortly before dawn"* Jesus came out to them, walking on the water. But this time he had no intention of calming the storm. It continued to rage. Now they really needed to know what they had to do not to drown. Peter called out, *"Lord if it's you, tell me to come to you on the water. 'Come' he said."*

Peter climbed out of the boat and walked on the water towards Jesus. But when he looked instead at the wind and waves he was afraid. Beginning to sink, he cried out, *"Lord, save me!"* Immediately, Jesus reached out, grabbed him by the hand and saved him.

THE ONES LEFT BEHIND

Our first two storms had come and gone. Charmaine herself had gone on ahead. We'd been together for almost 50 years. Then I was *The One Left Behind*, alone in this storm without end. I was alone, no longer even in the boat.

Jesus had beckoned me on towards him. Face to face, in eye contact with him, I walked on the waves - because he said so. Many times over the years since then Jesus has grabbed me by the hand as I looked down and was about to sink. But, again in the words of the song, *"Walking on water is just the beginning."* (Bethel Music) Nowadays, having reached the shore, I must learn to walk again on the land as best I can.

Charmaine was always with me, waking or sleeping she was present in my life, just as our parents, spouses, kindred and even closest friends are *present* in our lives. I only learned how constant and powerful her presence was when it stopped as she was no longer there. Punctum.

In a sense, not accidentally to paraphrase the Bible, *"in each other, we lived and moved and had our being."* (Acts 17:28)

Likewise, this illustrates to me how Jesus is always here with me, Emmanuel *"God with us."* For better and for worse we're rarely as full on together as before. And yet the Lord has somehow begun to fill that empty void left by Charmaine's departure. Who else could give the wise and faithful council I missed so much, knowing me as well as she did? Dare I say I almost miss the emotion of those times of intense grief when face to face with His presence?

Others have written of a similar experience where the greater the need the greater the sense of God's presence. Chinese Christians, released from many years of solitary confinement for their faith, have recounted this to Open Doors through its global ministry to persecuted Christians. God's presence was so close and so profound they almost regretted being unable to return to their prison cells to recover that intimacy with Him after their release. Likewise, with friends of ours;

the wife whose young husband suffered for many years from multiple sclerosis to the extent at one point he could barely get out of bed and began to believe his life was over. They recounted to me the same sort of experience. He was completely healed through prayer where medical treatment failed. He now works in a high-powered senior management role, but they almost miss those times when Jesus's presence during the worst times was all but tangible and saw them through. Even though the final outcome was different I can attest to that myself.

THE ONES LEFT BEHIND

Chapter 9

Conclusions So far

You taught my feet to dance on disappointment
And I, I will worship you
"Heroes" Bethel Music

What then are we to make of such turbulent events in our lives? For sure God does not cause them. Again, I live by a couple of *rules* put like this.

Rule #1: God is Good all the time - and He is Wise
Rule #2: If it ever appears God is not Good and Wise,
see Rule #1

In his book, *My Utmost for His Highest (August 24th)*[19] God gives what Oswald Chambers refers to as *'providential permission'*. This is really quite radical teaching but I have found it explains so much about how we respond to circumstances that seem to defy all Godly logic. God is after all absolutely in charge. He is Sovereign, King of kings. But in His wisdom, He might permit now only later to provide. He knows the end from the beginning. Not only does He never leave us, He also engages with us in a progressive, totally bespoke lifelong way.

He leads us towards maturity, so we are complete, not lacking anything. (James 1:4)

It's so easy to engage in scriptural tennis. It has been said, with some justification, '*You can use the Bible to prove anything you like.*' Arguments and counter arguments can go endlessly back and forth quoting one apparently contradictory scripture after another. It has also been said, '*Scripture quoted out of context is a pretext.*' Rather I see scripture as God's word which, when infused by the Holy Spirit brings us that, "*Peace that surpasses understanding with which we can safeguard our hearts and our minds*" from the destructive aspects of grief and anxiety. (Philippians 4:7)

Here is some hard teaching,

> "*Consider it pure joy, my brothers and sisters, whenever you face trials of many kinds, because you know the testing of your faith produces perseverance. Let perseverance finish its work so you may be mature and complete, not lacking anything.*" *(James 1:2-4)*

> "*Glory in our sufferings because we know that suffering produces perseverance; perseverance, character; and character, hope. And hope does not put us to shame, because God's love has been poured out into our hearts through the Holy Spirit.*" *(Romans 5:3-5)*

Hard indeed! 'How in the world am I to do that? You have no idea what's happened to me and how I've suffered!' one may say. That's so true, and I can but weakly sympathise. All I know for sure from my own experience, no more tragic than many of my readers' is, Jesus knows. So, we need have no fear through God's assurance for;

> "*I have redeemed you; I have summoned you by name; you are mine. When you pass through the waters, I will be with you; and when you pass through the rivers, they will not*

sweep over you..... When you walk through the fire, you will not be burned; the flames will not set you ablaze. (Isaiah 43:2)

What promises there are for us, not that we will not have to "*pass through rivers*" or "*walk through fire*" but however we may describe our circumstances, we can rest assured in God's almost incredible but true promise:

"I will go before you and make the crooked places straight; I will break in pieces the gates of bronze and cut the bars of iron. I will give you the treasures of darkness and hidden riches of secret places, that you may know that I, the LORD will call you by your name." (Isaiah 45:2 NKJV)

The following is not just for Simon and the disciples, but Jesus addresses us too in this hard teaching;

"Simon, Simon, Satan has asked to sift all of you as wheat. But I have prayed for you, Simon, that your faith may not fail. And when you have turned back, strengthen your brothers." (Luke 22:31)

Note carefully, Jesus's prayer is not that Peter not be *sifted*. His prayer is our faith should not fail, for us thereby to be strengthened and return to strengthen others. Deep stuff. I know. But when we're deeply mired in grief or bereavement we need somehow to gain traction to get out of it. We need hope, that in some way it might turn out to have some ultimate value, at least for someone.

Nevertheless, Charmaine and I did ask ourselves, Job-like from time to time, the kind of *Why us?* question. Each time she was reminded of a scene in the 1964 film *Zulu*, which you may have seen. A small unit of 150 British soldiers, thirty of whom were sick or wounded, is cornered in a military hospital at Rorke's Drift in South Africa's Transvaal back in 1879. They're surrounded by a highly disciplined force of 4,000 Zulu warriors intent on their destruction. The Zulus advance on the tiny outpost, protected only by sandbags, threatened with earth shaking menace. The Zulu war chant is made all the more intimidating by the thunderous sound of 4,000 spears impacting 4,000 raw-hide shields and the compelling pounding of 8,000 feet.

A simple Private soldier is pictured beside a towering, battle-matured, glitteringly decorated Colour Sergeant-Major. Terrified for his life by the sight of the advancing tide he tremulously asks, as well one might, as well we occasionally do, *"Why us Sarge? Why does it have to be us?"* With sonorous calm and aplomb, the bemedaled Colour Sergeant-Major's lips barely move as he intones aside, *"Cos we're 'ere lad."*[29]

This is not from the Bible, of course, but somehow God also speaks to us through film clips and so much more. This clip from Zulu gave us peace through understanding. Which tells us something wonderful about God. He does not restrict himself to his Word and will never contradict it. His Word is never wrong, but our interpretations of it most certainly can be. So, what are we to make of all this? His Grace was and still is sufficient.

Chapter 10

None Left Behind

*The Lord is not slow in keeping his promise,
as some understand slowness. Instead he is patient
with you, not wanting anyone to perish, but everyone
to come to repentance*
(2 Peter 3:7-9)

By nature, I'm curious. I still want to understand God's ways. I'll turn later to 'His means'.

But, why so soon? Why was Charmaine the same age as my mother and father when they too went on ahead through cancer? Why now? If it was God's will, then what more is there of his will left for me?

Mystery has its place, and we are not God to know everything. I love it when Moses asked God what he should say to Pharaoh when asked, "*Who sent you?*" "God said to Moses, '*I AM WHO I AM*' *Just tell him I AM has sent you*". (Exodus 3:14) This says to me God is God, uncontainable, unknowable by scientific definition. Not a very satisfactory answer, but in

His power and majesty, the Creator of all things - enough.

And yet he wants to be known and can be known by us. *"You will seek me and find me when you seek me with all your heart. I will be found by you."* (Jeremiah 29:13,14) But God is no push-over. In *The Chronicles of Narnia*, C. S. Lewis makes the most skilful use of sometimes regrettable circumstances to draw our attention to God's ways to who he really is, through Aslan.[30] As the French put it, *"Il est bon mais il n'est pas gentil."* God is Good - all the time: He is not always nice.

God speaks to each one of us directly when he says, *"As the heavens are higher than the earth, so are my ways higher than your ways and my thoughts than your thoughts."* (Isaiah 55:9) And yet it is, *"He who forms the mountains, who creates the wind, and reveals His thoughts to man - He who turns dawn to darkness and treads the heights of the earth - the Lord God Almighty is His name."* (Amos 4:13)

He indeed is the same One who so loved the world He gave us His son... (John 3:16) so we can know this about God, *"'I know the thoughts I have for you,' declares the Lord, 'thoughts to prosper you and not to harm you, thoughts to give you hope and a future.'"* (Jeremiah 29:11 NIV & NKJV)

It has been said, *"We look to God to change these 'sometimes regrettable circumstances', while he looks to them to change our hearts."* Either He is in charge, or He is not.

Life and Life to the Full

Within the breadth and the depth of all the challenges there are in life, Jesus came here for us to *"have life, and life to the full"* (John 10:10) - in the here and now. Our life in this world is full of all kinds of things, not all good. But they can't add up to true fulness of life without him being with us through them all. Jesus went to the cross to underwrite this promise, *"Never will I leave you; nor will I forsake you"* for *"Surely I am with you always, to the very end of the age."* (Hebrews 13:5, Matthew 28:20) It's when the Holy Spirit makes these words come alive in your hearts when they're most needed, is when all that can be said is, *"Yes. I believe!"*

Of course, I don't have all the right answers, but hopefully address some of the right questions. I don't know where I'm going, but do I know who I'm going with. I don't know what my future holds, but I know who holds my future. Also, I have this assurance of God's goodness and kindness no matter what.

I never refer to Charmaine's *death* but to her passing, her going on ahead. The cancer died, but she lives.

I need now to ensure my motivation doesn't fail. Now there are no daily words of love, affection or encouragement. There's no touch in the night, no letters, no texts or emails, no future meeting date to look forward to, no more *"see you sooons"* as always before. There'll be no more of this before my departure from this world, which does have its newfound attractions. Where is the *Off* button, so to speak?

But who am I not to follow through on what Charmaine had so faithfully endorsed, at what cost for so long, and as the Lord continues to lead me into fullness of life.

Since then, I have been better provided for, financially and materially, in ways I never have before in my life. Relationally everything changed. Charmaine went on ahead. My daughter, her husband and their four children (all under ten at the time) went to Canada and started a new life there. I moved away from my church family in Warwick back to my hometown of Manchester.

So, it is when key relationships in our lives are *"stripped away, when all I have is you, Jesus; I will pursue you"* in the words of the song, *Pursue*.[31] As Corrie ten Boom put it, found written recently in a little country church in Edale,

Derbyshire as if meant for me, *"All you need is Jesus until Jesus is all you have."*

And so we prepare for *the 'more'* stored up for us for wherever and whenever we are ready to receive it without harm, for *"He has gone into my future to prepare the way, and in kindness he follows behind me to spare me from harm from my past."* (Psalm 139:5 TPT)

Well, in fairness I can see where a purely scientific mindset can really struggle with topics like this, just as I did myself for the first two-thirds of my life. The natural human tendency is of course to want to understand in order to believe.

Jesus said many times, *"Fear not. Only believe"* as in do not fear getting it wrong. Psychologists would attribute this to breaking free from *"cognitive bias"* we are so prone to where we, *"create our own 'subjective reality' from our perception of the input."* Otherwise known as *"cognitive stereotyping"* we have a human tendency to happily accept information that appears to confirm an established *"bias"* or *"stereotype"* and more or less forcefully reject information that does not.[32]

My purpose here is to recount and share some of the careful and ongoing examination of my own belief system at a time of great challenge in my life. Thus far, I still revert to Jesus's words, *"Fear not"* as we are quite prone to do, *"Only believe."* that we may understand, not understand that we may believe. *"Not very helpful"* some may say. Well, let's move on and see where we get to.

Since the Lord, in His wisdom, *elected* for Charmaine to go home early, He must have something different for me to do than if she had remained.

I wanted life for her as her *champion,* as she sometimes called me. I was her *knight in shining armour.* I went out into life carrying her *colours* far and wide around the world, from China, through India and Africa to the US and back. I nevertheless feel disappointed sometimes in not having been there when I see how much time Charmaine spent alone embroidering so many cushions so beautifully, while I was so often away *championing*.

As I heard earlier at the Jordan, the Lord somehow has other plans as with, "*Let her go now, she's mine.*" She now has a new Husband. I submit to Him - and not without reluctance, I can tell you, but He is Good - all the time; He is God. All is not always well with my soul, but I tell it in the words of a song referred to earlier, "*Let go my soul, and trust in Him.*"

When I think of all the operations, surgical interventions (tubes into the aorta, under the skin, the vocal chords treatments, hospital visits, doctors' visits, tablets of all kinds, body creams, eye creams, mouthwashes, oxygen, wheelchair, hospital bed at home - I am just totally in awe of Charmaine's faithful, unwavering dedication to the process. She sometimes said she could not have done it without me. But although I hardly knew it, I drew much of my own strength from her inspired steadfastness. I long to tell her so. Life to the full? Yes, here and now with Jesus.

Life Here and Hereafter

Some things must remain a mystery for our own good. Are we yet fully prepared to "*know even as we are fully known.*" (1 Corinthians 13:12) But as an engineer I don't actually like mystery. I always like to know, to have explanations for why things are as they are. Who are we? Why are we here? Where are we going? I still have these restless questions which I will explore scientifically in chapters to come.

Fortunately, God still speaks to us today, directly and through drawing our attention to particular Bible verses. He wants to throw light on at least a few of these mysteries through his "*spirit of wisdom and revelation*" (Ephesians 1:17) so we can know Him better. Yes, He wants to be known by us, his creation, and why not? This is why Jesus came here at all, to reveal the Father, *"No one knows the Son except the Father, and those to whom the Son chooses to reveal him."* (Matthew 11:27)

So, who are those people by the way? Who does Jesus choose to reveal himself to and who does he choose not to - if any? Back to the user-manual and Jesus' own words in prayer at the last supper, hours before his arrest and crucifixion. *"Father, the hour has come. Glorify your Son, that your Son may glorify you. For you granted him authority over all people that he might give eternal life to all those you have given him."* (John 17:1-2)

This is going to be a hot topic, but I think it's really important to try to understand what Jesus is saying and praying here, to *"give eternal life to all he was given"* (John 17:1-2) It is after all part of the very foundation of Christianity. By implication were there some who were not *"given"* to him. Will some be left behind to go to hell? How could this ever be? Who can they be? In John 17, verse 9, Jesus does say, *"I am not praying for the world,* (not for everyone in the world) *but for those you have given me, for they are yours."* Once again, will some be left behind? But in the next verse he hears back in prayer and clarifies this by saying, perhaps with a tone tinged in extremis with acceptance and gratitude, *"All I have is yours, and **all you have is mine.**"*

I think the discussion can be concluded with, *"The Father loves the Son and has placed everything in his hands."* (John 3:35) Having resolved this matter John does complicate things in the very next verse (36) by saying, *"Whoever believes in the Son has eternal life, but whoever rejects the Son will not see life, for God's wrath remains on them."* What follows is no rabbit hole but a matter of where we spend eternity, in heaven or hell.

It's not for me to offer any more than my own view. For me, '*all*' means **all**. So, I struggle to believe God the Father would somehow choose to hold some people back, not giving them to God, the Son (i.e. God Himself), to die for. Even before the beginning of time, when all this was planned and foreseen, would God really consign people to hell before they were even born - from forever to forever?

Many of those who are told about Jesus simply choose not to accept who he is for this very reason - as did I for nearly 50 years. But again in Jesus' own words, "*Come to me **all you who are weary and burdened** and I will give you rest. Take my yoke upon you and learn from me, for I am gentle and humble in heart, and you will find rest for your souls. For my yoke is easy and my burden is light.*" (Matthew 11:28)

Our only qualification to be *called* is at some point in life to feel weary and burdened. Is anybody feeling left out? Still, I hear the anguish in the cry, "*How come a faithful, merciful God who IS Love lets people burn in hell - forever?*"

Jesus was nailed naked to the cross and there was tortured to death so nobody would need to go to hell. This does not mean to say hell does not exist or nobody will ever go there. Back again to the *user guide*, the Bible.

Hell was made for the devil not for man. (Matthew 25:41) In Romans 8:29 Paul assures us, "*Those God foreknew he also predestined to be conformed to the image of his Son.*" God could not fail to foreknow all those he created and therefore predestined. None of them should therefore be lost. All should be saved somehow, to be conformed to the image of Jesus - with none left behind.

Peter did say "*The present heavens and earth are reserved for fire, being kept for the day of judgment and destruction of the ungodly*" but went on to explain, "*Do not forget this one thing, dear friends: With the Lord a day is like a thousand years, and a thousand years are like a day."* And concluded with *"The Lord is not slow in keeping his promise, as some understand slowness. Instead **he is patient with you, not wanting anyone to perish, but everyone to come to repentance**"* (2 Peter 3:7-9) at some point in time.

However, a core Christian belief is we can either enjoy eternal life with Jesus in heaven or burn in eternal agony with Satan in hell. It's our choice. God seems to prioritise our freedom to choose one or the other over his own desire for none to perish.

And yet, *"He has saved us* (past tense) *and called us to a holy life—not because of anything we have done but because of his own purpose and grace. **This grace was given us in Christ Jesus before the beginning of time.**"* (2 Timothy 1:9) We were all there on his mind even before the beginning of time. As they say, "You can't make this stuff up!"

Fortunately Jesus promises, *"I will draw all people to myself."* (John 12:32) Through his "kindness" not fear of hell, God will *"lead us to repentance."* (Romans 2:4) Christianity traditionally stresses our need to exercise this freedom of choice before we breath our last. To go to heaven we have to say and believe "the sinners prayer" or otherwise "come to faith" before that deadline or we'll end up in hell and **the deal is off**.

After that Jesus apparently stops drawing us to him. His patience and kindness come to an apparent end. No longer led to repentance, we're on our way to hell. Even as our spirit has its first sight of the fires of hell, at the first whiff of sulphur, there's no going back. Too late! We don't usually hear it put this way, but it is a core belief of Christianity.

How then can this be? How can a faithful, merciful God who IS Love set things up with a tangible option for people to burn in hell in agony for having made the wrong choice - forever?

No longer driven by Dante's own grotesque visualisations of hell, those who don't believe in heaven don't believe in hell either. Choice becomes a moot point if there's no clear picture of both from which to choose. Once you do believe in heaven only then do you get to believe in hell - after which you can be pretty sure you're not going there anyway.

We may well ask, "**Who then can be saved?**" Jesus assures us, "*With man this is impossible, but not with God; all things are possible with God.*" (Mark 10:25-27)

We know that our spirit is eternal, bound for eternity either in heaven or in hell. Yet we still have a way of assuming death of the body is the end, a kind of last chance saloon. Yes, death is a life-changing experience, but it's not the end. Although for us at that point we finally lose all our cherished control over ourselves and over others, for God it's merely a change of venue.

I'd suggest instead it's NOT IF someone actually gets to make a fully informed choice BUT WHEN. It may be as late as the arbitrary moment of death of our body - or even later for our eternal spirit.

But the Bible does say, "*At the name of Jesus every knee will bow and every tongue confess...*" (Philippians 2:10-11) What does this really mean if not "every?" Might non-believers be forced at last by some violent means to submit and "bow the knee" as they die before being sent immediately to hell? How might that look? How long would anyone need to spend on a quick guided tour of hell to realise it's not for them, even if they ever volunteered to take the tour? Seriously!

But it does say "At the name of Jesus" meaning when they hear Jesus himself speak his name to them? Many sovereignly come to faith that way in this life even now.

Or imagine instead those final moments when Jesus comes back for **all** for whom he promised to prepare a place (John

14:3) face-to-face with God who IS love, eyes locked with those of Jesus. He paid **once for all**, i.e. for everyone, with a "love as strong as death" (Song of Songs 8:6) Everyone got died for. Now he can't wait to see at last all those who were "the joy that was set before him to endure the cross" (Hebrews 12:2) for them to spend eternity with him instead of like "the rich" man reaching out to Lazarus from hell. (Luke 16:24)

Who will not exclaim when they see Jesus, "*Look! Here he comes, leaping across the mountains, bounding over the hills*" (Song of Songs 2:8) to them and not say *Yes!* to him? I have no doubt this is how he came to Charmaine on such a dark winter's night. I could all but sense the two of them just having skipped off together as I came downstairs. Her *winter* had passed, and her springtime had come. (Song of Songs 2:11-12)

So, the mainstream belief is that unbelievers are despatched immediately to hell the moment they die. If they neglected to make a choice, or made the wrong choice beforehand, they failed to reserve their place in heaven, so to speak. But, the Bible says, "*Just as people are destined to die* (only) *once, and* (only) *after that to face judgment.*" (Hebrews 9:27)

If some came to believe in Jesus in this life without seeing him, will not the rest believe when they do, especially after they just died and be saved - even atheists? "*The one who believes in me will live, even though they die.*" (John 11:25)

Picture then the scene at this new venue. How attractive then the fires of hell as the devil is hauled off to them in chains? Who then will follow the smoking tail of the one they now recognise as the lifelong abuser of their souls. Will they really choose to join him in hell, to burn in agony with him there forever?

What then will a person do? As each one shows up, the "prodigal Father" will run over to greet them as the individuals Jesus died for. He became sin itself so we could

all become the "righteousness of God." This he did while we were **all** sinners, non-believers **all**, before we were even born let alone before we died. Who then will choose to deny the love of God right there before them? Surely every knee will bow in awe, and every tongue will joyfully confess, "Jesus is Lord." (Philippians 2:11 NKJV)

Even so, despite what happens "at the end" it's sad for those who have to manage all the way through life without experiencing the love of Jesus. They will have lived a "morning after the night before life" instead of a *"life and life to the full"* kind of life. (John 10:10) Never do they get to minister the love of Jesus to anyone. Nor do they graduate from the "boot camp" of this life, not having entered training for reigning in the next one, not having begun to learn to love unconditionally as Jesus loves.

Speaking of himself Jesus explains, *"For judgment I have come into this world..."* (John 9:39) Addressing the church Paul asserts, *"We must **all** appear before the judgment seat of Christ, so that **each of us** may receive what is due to us for the things done while in the body, whether good or bad."* (2 Corinthians 5:10)

Furthermore, those who do not believe in Jesus cannot tell another soul the good news about him. So, it's better to have the greatest number in the boat, so to speak. Heaven knows it's big enough. They in turn can, as the numbers grow, bring others aboard during their lifetime not just at the last minute.

Here again, this is where I can almost hear some readers joining others in reaching for the stones...

Even so, might I urge we carefully reflect on what Jesus means when he says, *"No one can come to me unless the Father who sent me draws them, and I will raise them up at the last day."* (John 6:44) Does he really mean the Father will draw some and not others? Does he really mean some he will lift up on the last day and some not?

Just to either confuse or complicate the subject, depending on which side of the debate you favour, but also to lighten it a bit, I love this story of Peter at the Pearly Gates.

Peter is there with his assistant. In the book of life they're checking people off as they enter heaven. Those whose names are not written in the book are despatched to hell. The assistant takes off into heaven to tally the numbers. Breathlessly he returns to complain to Peter. "You know boss, according to our numbers there are way more people in there than we've ever let in." "Hmm. Tell me about it," shrugs Peter with a big sigh. "You're new here, aren't you? But it's always the same you know. It's Jesus! He keeps letting them in over the back wall!"

But seriously, consider this. Whoever in this world does not recognise and accept Jesus for who he is will have to live here through all the trials and tribulations of this life without ever knowing his goodness and comfort - until they do. I can join everyone who speaks from experience of a time before coming to faith, of just how hard it was without Jesus and although not perfect, how much better it's been since. Jesus longs to be with us all through it all.

If that really is true for us all, for "**each of us**" as believers and unbelievers, how then are we to understand, "*Mercy triumphs over Judgment?*" (James 2:13) **Is it true or does God actually stop being merciful after an unbeliever's body happens to die** before they get it right? It's not as though God is setting an example to those yet to come since nobody will really know much about hell until they get there anyway.

Let us recall, "*In him we have redemption through his blood, the forgiveness of sins, in accordance with the riches of God's grace.*" (Ephesians 1:7) Do the "*riches of God's grace*" really run out for non-believers just as they die? "*He sacrificed for their sins **once for all** when he offered himself. Made holy* *through the sacrifice of the body of Jesus Christ **once for all**.*" (Hebrews 7:27, 9:26) Does that "**all**" really mean **all**, or only some who made the right choice before they died? How else are we to read "all" in, "*As all die, so in Christ all will be made alive*" (1 Corinthians 15:10) being assured a few verses later that, "*The last enemy to be destroyed is death?*" (1 Corinthians 15:10) Both statements are all inclusive. Neither one is contingent on our believing this before we die.

So imagine the scene. After passing judgment, again on us all for, "*...what is due to us for the things done while in the body whether good or bad*" (2 Cointhians 5:10) Jesus steps down from the bench and declares, "It is finished! They're with me! I've paid the price. Case dismissed!"

Can it really be as simple as that? "The last should not be let into heaven at all after they die!" we might complain. "It's not fair!" But who are we to say the last should not be first and God does not have the right to be generous with his salvation? (Mark 20:1-16) "But surely there's something we should do to deserve not going to hell? Anyway what about...?" We can fill in our own blanks.

And yet, even if it remains a matter of personal choice, we can of course insist on being condemned, refuse to have our case dismissed and purposefully head off instead to hell. But, is there not a strange inconsistency here with the faithful merciful God we know, who IS love? How did Hell, created specifically for Satan and all his demons, actually become the alternative destination for the world he so loved? Most certainly, "*The wages of sin is death*" (Proverbs 10:16) but is that not the reason why Jesus came here at all? For, "*the **Gift** of God is eternal life in Christ Jesus our Lord.*" (Romans 6:23) So now it's "*Death where is your sting?*" (1 Corinthians 15:55)

Should the C. S. Lewis we love have the last word? In "The Great Divorce?" he claims, "There are but two kinds of people in the end. There are those who say to God, 'Let thy will be done' and there those to whom God says in the end, 'Let thy will be done.'"

Would Jesus really have come here to be crucified just so our free will to choose to go to hell could be superior in a sense to God's will for us not to? E.g. 2 Peter 3:9, Romans 8:29 and more. Or will Jesus do whatever it takes even after we say farewell to our bodies, to be there to be sure none of us make the wrong choice even by accident?

Now of course, Jesus speaks of hell in the New Testament more than it is ever mentioned in the Old Testament. He wants not one single person to be at risk of going there at the last moment by default. He, better than anyone, knows what it's like. He himself as God created hell for the devil, not as a principal residence to call his headquarters, but as a place of eternal banishment. Jesus even went down there himself, as recounted between his crucifixion and resurrection, to *"the depths of the earth."* (Ephesians 4:9) Obviously, this is neither the earth's geological molten mantel, nor limited in time for an earthly day or two. Could it be Jesus timelessly checks to see if anyone who happens to find their way down there is not left behind?

What Then are We Here For?

At this point it would be remiss of me not to rephrase the question, speaking of all these trials and tribulations of life happening to everyone, what's the point? *"What are we here for anyway?" "What's it all about?"* as Alfie put it, speaking directly to the camera in the film of that name.[33] What indeed?

If life ever feels like a kind of boot camp, in training to be Marines, Navy SEALs or SAS agents to survive the deprivations of some remote desert, ocean or jungle location, then maybe we are. Look at the world around us. Are we all perhaps in training to *"reign in life"* (Romans 5:17) to be at the wheel of life with Jesus instead of under its wheels without him. Maybe we are?

What if we are *'training for reigning'* as Leef Hetland puts it?[34] Training for reigning in life here AND in the 'hereafter' somewhere in the 20 times more undetectable universe which co-exists with us in the part we see. It was mentioned earlier and we will revisit it in some detail. Astrophysicists refer to this as the 'dark matter' and 'dark energy' we hear so much of nowadays.[35]

We all know just how bad it really can be here on earth. I have to believe it is just as good in equal measure in heaven - in the other 95% of the universe, of which Science can tell us next to nothing apart from it being here and now amongst us. It is *"at hand"* and even *"within us"* as the Kingdom of Heaven (Luke 17:21 NIV and NKJV).

Now, we don't want to be "so heaven focused we're no earthly good", as they say. Neither do we want to be so earthly focused to be of no heavenly good.

I reflect sometimes on the story of Steve Jobs' passing, a committed atheist all his life, as recounted by one of his sisters.

Steve died at his Palo Alto home around 3pm on October 5th, 2011, due to complications from a relapse of his previously treated pancreatic tumour which resulted in respiratory arrest. He had lost consciousness the day before and died with his wife, children, and sisters at his side. His sister, Mona Simpson described his death thus: *"Steve's final words, hours earlier, were monosyllables, repeated three times. Before embarking, he'd looked at his sister Patty, then for a long time at his children, then at his life partner, Laurene, and then over their shoulders past them. Steve's final words were, 'Oh wow! Oh wow! Oh wow!'"* He then lost consciousness and died several hours later."[36]

Nobody can say of course, but might this have been the Jesus he saw?

> "Dressed in a robe reaching down to his feet and with a golden sash around his chest. The hair on his head was white like wool, as white as snow, and his eyes were like blazing fire. His feet were like bronze glowing in a furnace, and his voice was like the sound of rushing waters."
> (Revelation 1:13-16)

Hardly surprising Steve would exclaim "*Oh Wow!*" so emphatically. Who then could ever say, "*No*" to this voice "*like the sound of rushing waters*" - to this Jesus?

It was only slightly different for atheist musician, actor and comedian Dudley Moore whose last words are reported to be, "I can hear the music all around me!".[37] "*For what they were not told, they will see, and what they have not heard, they will understand?*" (Isaiah 52:15)

But then, of course there's the question of the Sheep and the Goats! (Matthew 25:31-46) Now we're getting somewhere.

Many of us say, '*I can't believe in God who IS would send people to burn in hell at all, never mind for-ever.*' Whether such a god exists or does not exist is not a matter of personal opinion. It would be a matter of fact, not of our belief. We don't get to choose what is. However, I don't believe in the same god they don't believe in. But I do think it's important to enquire where this view comes from of a kind of, "*Saddam Hussein in the sky who keeps torturing people all the time,*" rather bluntly expressed as the views of some.

One of the most stark and apparently definitive is the well-known parable of the sheep and the goats. Given its significance it seems sensible to read it in full in Matthew 25:31-46 (the emphasis is mine).

> "When the Son of Man comes in his glory, and all the angels with him, he will sit on his glorious throne. All the nations will be gathered before him, and He will separate the people

CHAPTER 10 - NONE LEFT BEHIND.

one from another.

As a shepherd separates the sheep from the goats, He will put the sheep on his right and the goats on his left. Then the King will say to those on his right, 'Come, you who are blessed by my Father; take your inheritance, the kingdom prepared for you since the creation of the world.

*For **I was hungry** and you gave **me** something to eat, **I was thirsty** and you gave **me** something to drink, **I was a stranger** and you **invited me** in, **I needed clothes** and you clothed **me**, **I was sick** and you looked after **me**, **I was in prison** and you came to visit **me**.'*

*Then the righteous will answer him, 'Lord, when did we see **you** hungry and feed **you**, or thirsty and give **you** something to drink? When did we see **you** a stranger and invite **you** in, or needing clothes and clothe **you**? When did we see **you** sick and in prison and go to visit **you**?'*

*The King will reply, **'Truly I tell you, whatever you did for (even) one of the least of these brothers and sisters of mine, you did (do) for me.'***

*Then he will say to those on his left, 'Depart from me, you who are cursed, into the eternal fire **prepared for the devil and his angels**.*

*For **I was hungry** and you gave **me** nothing to eat, **I was thirsty** and you gave **me** nothing to drink, **I was a stranger** and you did not invite **me** in, **I needed clothes** and you did not clothe **me**, **I was sick and in prison** and you did not look after **me**.' They also will answer, 'Lord, when did we see **you** hungry or thirsty or a stranger or needing clothes or sick or in prison, and did not help **you**?' He will reply, 'Truly I tell you, **whatever you did not do for one of the least of these, you did not do for me.'***

"Then they will go away to eternal punishment, but the righteous to eternal life."

How well, may I ask of myself and each of us, would we as believers come out in this tally separating sheep from goats? Are we really keeping count? Anyway, what must the requisite number of such deeds be in order not to count as a goat (going to eternal punishment) but as a sheep (going to eternal life?)

Might we be missing something here? What is Jesus really telling us? Who is this really about after all? Looking back over the bolded words, the focus is not on personal performance. It's all about Jesus and how we minister Him to others. If we make Jesus #1, not ourselves and our position on a kind of spectrum of performance-based salvation (itself a contradiction in terms) there is no #2.

And yet, we have this terrible human compulsion to measure our own performance and compare it with the performance of others. In this case, it's what we might believe to be God's measure of approved or sanctified behaviour? Again, to how many needy people do we actually have to give food or water to qualify as a sheep or a goat? Should we not be doing that anyway as we "love our neighbour as ourselves?" But I question should we, can we ever, keep count and compare performance to an unwritten sheep or goat target number?

For, *"It is by grace you have been saved, through faith— and this* (faith) *is not from yourselves, it is the gift of God not by works,* (such good deeds) *so that no one can boast"* (Ephesians 2:8) of being a sheep and not a goat.

Likewise, if and how many strangers have we invited into our home, clothed them or visited them in prison? Even if we could count them all over our lifetime what are the threshold levels for each category to qualify to be counted either as a sheep or a goat? How could we ever meet or even exceed some kind of unquantified spiritual Olympic standard?

Needless to say, this could be serious. Do we qualify to go to be blessed with an *"inheritance in heaven"* or *"cursed, into the eternal fire prepared for the devil and his angels?"* Note here for whom hell was actually prepared.

CHAPTER 10 - NONE LEFT BEHIND.

It's quite natural to want to know how many of each we have to do of course. But this is after all a parable and with some of Jesus' characteristic hyperbole. (Matthew 18:8)

However unquantifiable, should it really be taken literally and is there a more important theme we might consider? Whatever it is, we'd better not get it wrong given the consequences, one might say.

And yet we know, *"God so loved the world that he gave his one and only Son, that whoever (and whenever) believes in him shall not perish but have eternal life."* My addition in brackets is inspired by the *"last will be first"* story of the owner of the vineyard who paid everyone the same regardless of what time of day they began work in Matthew 20:1-16. Here it's all about *"whoever believes"* not *"whoever provides food, water, hospitality and clothing to those in need or visits them in prison"* at whatever time or in whatever quantities.

What I'm looking at here is no theological technicality. In context it's a matter of eternal life or eternal worse-than-death. It involves a direct contradiction between belief and a whole undefined series of actions, for a whole variety of people over an entire lifetime. Any shortfall in any of which could catch us out - forever.

Non-believers intuitively see this contradiction which defaults to the, *'I can't believe in a god who would send people to burn in hell.'* Does this really make sense from what we know of Jesus, and if not, what actually is his point?

It was he who answered this question *"Of all the commandments, which is the most important?"* by saying *"The most important one is this: 'Love the Lord your God with all your heart and with all your soul and with all your **mind** and with all your strength.' The second is this: Love your neighbour as yourself.' There is no commandment greater than these."* (Mark 12 29:31)

These commandments are required to manifest as a state of mind, a manner of being, not a scorecard, as some may see it.

I would suggest this parable is not about who goes to heaven or hell. That is *"by faith alone, by grace alone"* not by *"works"* or what we do by which we are saved from it. (Romans 3 & 4 and Ephesians 2:8,9) This indeed is a pivot-point of Protestant Christianity.

I would suggest instead this parable is more about personifying Jesus in the people we offer any, and all, service to. It emphasises many times over, whatever we give or do we are actually giving and doing to Jesus. **"Whatever you did for (even) one of the least of these brothers and sisters of mine, you did (do) for me."**

It's not just about the quantity but also the quality of what we do. So, let's do it well, with the heart-attitude and the love for them as we have for Jesus himself. This was exactly Mother Teresa's heart attitude. *"Minister to the those who are hungry, thirsty, to strangers needing clothes, who are sick or in prison as if they were Jesus"* - because according to her, they are. Heaven knows we need that kind of encouragement sometimes?

Nevertheless, even the so-called *"righteous,"* referred to in this passage as its apparent beneficiaries, still didn't get it because they asked, *"When did we ever see you...?"* Verses 37-38.

I'm reminded of a stage performance I once saw in which a couple are anxiously tidying up and arranging their home. Every so often one anxiously chivvies up the other, *"Come on! Hurry up! Jesus will be here at any minute."* Of course, everything has to be just right for Jesus when he comes to visit them. At last, everything is perfect, the kettle's on, the scones are still warm.

As they sit fidgeting by the door, they hear a feeble knock. They stumble over each other in their haste to answer it only to find there a street beggar dressed in rags, holding out a begging bowl in filthy, encrusted hands. Of course, they turn him brusquely away and slam the door shut in his face. *"How dare he come at such a time as this when we're expecting Jesus at any minute!? Who does he think he is?"* they complain to each other.

So, they hurriedly dust themselves down from whatever contamination may have been conveyed during the brief encounter. They sit once again, fretfully waiting. They wait, and they wait - and wait until the scene eventually ends. Jesus apparently never does show up. The closing line?

"Anyway, who did he think he was anyway? That beggar spoiled everything!"

Oswald Chambers puts it this way in My Utmost for His Highest *"We are here not to save souls but to disciple them."*[38] (Matthew 28:19) Salvation and sanctification are the work of God's grace not of our performance in the competition to become a sheep instead of a goat.

As it is in Heaven

Meanwhile when one day I do see Charmaine again, I can almost hear her say to me when there I awake, *"Hello Mike! It's me, Charmaine, you're home."*

I know I have been left behind here to fulfil a legacy, to join with others to bring heaven to earth in some ways. Revealed to me in recent years has been the wonder and the power of God's love. Not only is this love manifested in and through us, mostly in ways which we are fortunately unaware. The lavishness and the very power of God's love is everywhere for us to see with our own eyes in the world around us - so magnificently evidenced by David Attenborough, Professor Brian Cox and many others

Can we meanwhile assert that **the blood of Jesus IS sufficient? He did not fail in His mission**. The 99 will be saved as he goes after the 1 so **in the end there are none left behind.** (Luke 15:4)

Chapter 11

The Power of God's Love

*Through the 'strong force' acting
on protons all things hold together
(Quantum Mechanics)*

More than any other verse in the Bible, John 3:16 is what most Christians know by heart, *"For God so loved the world he gave his one and only Son..."* Some may be less familiar with the rest of the verse, *"...whoever believes in him shall not perish but have eternal life."*

Although delighted to hear we are not somehow to perish, understanding and believing in the *eternal life* part can be quite problematic. *"Whoever believes"* is promised to have such life - seemingly for ever! Fortunately, we don't actually need to understand to have faith since *"the peace of God surpasses all understanding..."* (Philippians 4:7 NKJV)

Not only so, but this peace will, *"guard your hearts and minds .... so, we need be anxious for nothing."* (Philippians 4:6-7), even though sometimes we are.

This is all very well, but faith in eternal life is such a huge subject and I will come back to it soon. Sufficient for now, *"the hope of eternal life which God, who does not lie, promised before the beginning of time."* (Titus 1:2) Yes, before the beginning of time, scientifically established to have been before time actually began 13.8 billion years ago - yes, before then. So, before returning to faith in eternal life, I'll just park the discussion for a little while with Eli's Zen-like definition of faith (spoken by Denzel Washington) from the movie *The Book of Eli*. *"Faith is knowing something when you don't know something."*

What we do *"know"* from the Bible is, *"God IS love."* He *"so loved the world he gave..."* As I keep saying, true not just because it's in the Bible but, in the Bible because it's true. Having a mindset of curiosity myself, and for anyone else who'd care to join me, I want to know more about this Love with which God so loved the world.

Now, the subject of God's love does raise profound emotional and theological questions for us all such as, *"Where was God when.....?" "Why did God allow.....?" "If God is both love and all-powerful then why did he not.....?"*

These anguished questions, in many peoples' minds, arose in mine too from Charmaine's cancer and her passing. They led me to Jesus's own words - always a good thing to check in times of doubt - *"The Son can do nothing by himself; he can do only what he sees his Father doing, because whatever the Father does the Son also does."* (John 5:19) From this we learned to focus on what God is doing and not to stumble over what he is not doing.

Back to basics with Jesus, *"In him all things were created: things in heaven and on earth, visible and invisible ...... all things have been created through him and for him."* (Colossians 1:16) What then are these invisible things? How

CHAPTER 11 - THE POWER OF GOD'S LOVE

in the world did he do it? How did he put it all together? And again, how does he hold it all together even now?

Which takes me forward once again to God who doesn't only do love. He IS love. He has no option but to be Himself - to be love. We also know, *"In Him all things hold together"* (Colossians 1:17) What things? All things! Everything! How then does God actually do it - hold all things together?

Even with a scientific mindset as well as a curious one I bumped over these verses many times without indulging my curiosity much further.

Looking ahead to Chapter 12 - The Kingdom of God is Within Us, we will delve into the three 'energy fields' pervading the whole universe from which all matter, everything - is made. They are the 'electron field' the 'quark field' and the 'Higgs Field.'[39] We will also look at the four 'fundamental forces' acting upon these three 'energy fields'. They are the 'electromagnetic force' the 'strong force' the 'weak force' and 'gravity.'

Beginning with the most familiar, 'gravity' is an exceptionally weak force. It's enough to make it hard for us to run up a flight of stairs and so on. It literally keeps our feet on or very close to the ground, more firmly for some of us than for others. But it literally takes a mass the size of our planet to create enough gravity to do so.

Gravity acts over intergalactic distances of billions of light years - throughout the cosmos. But there are equally profound forces working only at the sub-atomic level, sometimes referred to as the 'particle-cosmos.' Not only are these 'energy fields' and 'forces' everywhere, they're in and through everything and everyone. They are what we're made of body, soul, (defined just for our purposes here as that which makes sense of our senses) thoughts and even our memories - although definitely not our spirit, not our *'spiritual operating system'* so to speak.

By way of context the 'nucleus' of an 'atom' is as small to a grain of sand as a grain of sand is to a cathedral. But in the terms of this illustration, it's heavier than the cathedral itself. As small as it is, almost all the mass of an atom is in the 'protons' and 'neutrons' making up the 'nucleus.' Then there are the 'electrons' of course, whizzing around the nucleus. In a rough-cut model of the atom, they can be compared with tiny mosquitos whizzing around the cathedral weighing about as much as mosquitos do compared with the weight of the cathedral.

Just to track back again to why I position Jesus in this world of science? Science quite simply describes His Creation.

Looking more closely at the positively charged 'protons' which are all bunched up together in the 'nucleus' of the 'atom.'. Most of us will have noticed when playing with two magnets, how the two North poles or two South poles repel each other. As we try to bring them closer together there's this funny squishy feeling showing us quite forcefully, they don't want to be pushed together. The closer they come together the bigger the force wanting to keep them apart. An invisible 'electromagnetic force' causes these "like poles" the same kind of poles of a magnet, to repel each other. They seem to want to stay apart.

The 'electromagnetic force' causes this phenomenon between magnets. It also acts upon 'protons.' Having positive "like charges" the 'electromagnetic force' acting on all 'protons' makes them repel each other very powerfully. This is especially so since inside the nucleus of the atom they're really very close together indeed. How close? Well, the size of a proton is 0.84 metres with 15 zeros after the decimal point (0.00000000000000084 metres). Each 'proton' is no further away from the next than this, very much closer than the distance between the like poles of our magnets. This huge electromagnetic force of repulsion must be overcome for them to still hold them together.

Nuclear physicists have named this rather prosaically the 'strong force.' Not only would all 'protons' and therefore all matter blow violently apart unless held together by the 'strong force,' nothing would ever have come together in the first place, "*visible or invisible.*" There would be no universe - and we would simply not be here either.

To give an idea of just how strong this 'strong force' is, "*holding all things together*" it's 10^{41} (as in 10 followed by 41 zeros or a full couple of lines of zeros) times greater than the 'force of gravity' which we know so well.

Since "*God is love*" (1 John 4:8) and "in Him all things hold together" (Colossians 1:17) I would suggest God's love also expresses itself as just exactly that - the 'strong force.'

Well, gee-whiz some may say. That's all very well, so neat. How do you know for sure it's true? Well, of course I don't know any more than I know it's not, which is where faith comes in, "*Knowing something when you don't know something.*" And if not, then what is going on here anyway? Doesn't it also seem somehow improbable to think all this is all here just because it is, we're just here because we are?

As Paul puts it in 1 Thessalonians 4:13, "*We want you to be quite certain about the truth* (of God's love) *concerning those who have passed away.*" It's important, "*so that you will not be overwhelmed by grief like many others who have no hope.*" (TPT)

Meanwhile just think, God's love is 10^{41} times the force of gravity! In this there is hope for someone like myself and anyone else who was left behind here to grieve the loss of the one who went ahead.

No wonder we need power even to know God's love as it's described here, so Paul prays, "*May you have power, together with all God's people, to grasp how wide and long and high and deep is His love and know this love that surpasses knowledge—that you may be filled to the measure of all the fullness of God."* (Ephesians 3:18-19) I think it's important to note what Paul is not saying here. He's not saying God's love replaces knowledge. God's love surpasses knowledge. It both surpasses and validates the scientific knowledge we have, incomplete and subject to review as it always has been. I remember my father quoting what is generally attributed to Aristotle, *"The more we know, the more we know we don't know."* What we do know is merely that the 'strong force' works.

Might the power of God's love for us be aptly illustrated once again in this way, by Eli's (Denzel Washington) statement of faith which is, *"Knowing something when you don't know something?"* Jesus declares, *"No one can snatch us out of my Father's hand"* (John 10:29) held there by the power of God's love.

Faced with this Biblical mystery might we contrast it with the scientific mystery of the power of the 'strong force' being 10^{41} times greater than the force of gravity, the very force of gravity with which the moon physically lifts up the tides of the oceans of the earth by many metres?

If this brings some understanding, know how much greater the peace can be which surpasses such understanding we do have. It can guard our hearts and minds, so much in need nowadays of being guarded they are. (Philippians 4:7) There is a backstop of faith. Right. A hard backstop of faith emerges as we explore more of the scientific and Biblical facts revealing the Kingdom of God both around and even within us.

Chapter 12

The Kingdom of God is Within Us - Scientific and Biblical Facts

To see a World in a Grain of Sand
And a Heaven in a wild Flower
Hold Infinity in the palm of your hand
And Eternity in an hour
(Willian Blake Auguries of Innocence)

Since the 1970s, astronomers have known galaxies have far too little matter at the outer reaches of their arms for them to rotate quite evenly right across their diameter without flying apart. These outer arm portions are observed to travel along progressively faster the further out from the center they are. The outer portions rotate in alignment with the inner portions. However, the further out they are the fewer stars there are in these arms and the less visible mass or matter they are seen to contain. Which means there should be less gravitational pull holding them in to stop them from flying off into space. But there's actually more gravitational pull holding them together than expected from the mass we can see. It's coming from somewhere; just enough so they do.

It's a bit like a carousel at the fair. The horses at its edge travel much faster than the centre where someone can stand and hardly be moving. In this example the whole thing is held together by steel struts so the horses don't all fly off with us on them.

Likewise, something is in this case concealed within the arms of galaxies, including our own Milky Way galaxy. It's holding them more tightly together by a bigger gravitational force than their mass should generate. The *something* has been identified as 'dark matter' - dark because it's invisible, it does not interact with light or any other electromagnetic radiation such as X-Rays or anything else. It interacts only with gravity. It is still therefore very much part of the same 'spacetime' we ourselves inhabit. It's a different kind of matter altogether from what we just discussed made from 'protons', 'neutrons' and 'electrons.' It's not that at all. Yet 'dark matter' adds about five times more to the mass of the known, familiar 'ordinary matter' of which stars and planets and we are made. Rather more on this later.

By the 1990s, we discovered the universe is not only expanding but doing so at an accelerating rate. Something is 'pushing' it not only to continue to expand but to do so faster and faster. The missing something has been identified as 'dark energy.'

To put all this in a context of where we fit into all this intergalactic mass and energy. As I write, I'm speeding along at 100,000 km/hour. Fortunately, you are effortlessly doing the same as you read this since we're both on the same planet. We need to keep this up to complete our journey of nearly 940,000,000 km around the sun every year. Just as our tiny globule of molten magma we call Earth does this

CHAPTER 12 - THE KINGDOM OF GOD IS WITHIN US

at 100,000 km/hour it also spins on its axis at close to the speed of sound at the equator. Yet, here we quite happily are.

Somehow, we don't seem to notice, but we always flash past the same point together in space at the same time every year - at 100,000 km/hour or over 82 times the speed of sound, assuming that to be 1,222 km/hour in air at sea level.

Oh, and just to complete the picture we, together with the rest of our solar system, also orbit a super-massive black hole, 4 million times heavier than our sun, at the centre of our Milky Way galaxy. This we do at 750,000 km/hour or 7.5 times the speed of sound. When added to this 82 times the speed of sound it means we're going through space at nearly 700 times the speed of sound - without even knowing it.[40] It creates not the flutter of a leaf or the tiniest ripple on open water.

Now I must say, as I look around my room and out of the window, like you I don't have the least impression this is really happening. Nevertheless, although things around us are extremely convincingly real as they are, things are not quite as they seem.

To locate ourselves on a different scale altogether, we each began life as just one single microscopic cell no bigger than the thickness of a human hair. At which point we were no more than a single-cell organism - albeit of a very special kind. Somehow, the one cell contained everything necessary to become the unique, never before or ever again person we each of us are. This one cell was a perfect blueprint of a grand design made of a modest variety of atoms which became the glorious array of billions of billions (not just billions and billions) of molecules we have become.

The point is, contrary to all the indications of the centrality and importance of the reality of our daily existence here on earth - earth doesn't really count for much in the scale of things. Yet we do have these most extraordinary features of our existence in common with all the rest of the universe. Furthermore, unlike any other living thing, we know it. We can reach out to see and engage with it, and even go out there. The Voyager programme has sent satellites to visit all the planets. However, it will be another 14,000 to 28,000 years before they properly leave the solar system even travelling at close to 40,000 miles per hour as they do.

I include this factoid and more as measures of our own significance in our own solar system let alone our own galaxy. To this I say with one breath, "How can there be a God who did all this yet who knows us and cares for us?" Then with the very next, "How can there be a God who did all this yet who can NOT know us and cares for us?" So far beyond our mental capacity it is to figure out if God exists or not. This is my "hard backstop to faith" on days I need one and don't flip between the two. But it always comes down to having done life with Jesus over the years until I "know in my knower" He's real, just as I know my name or what's day or night, neither less nor more certain than that.

Returning to what we and the universe are made of, there are only 92 different naturally occurring kinds of atoms or 'elements'. They're the building blocks from which our bodies, minds and even our most distant memories are made. Each atom of each of the 92 kinds, is perfectly identical to all the others of its kind - throughout all the intergalactic vastnesses of the universe. Everyone and everything is made from the same 'atoms' of these same 'elements' - whoever was, whoever is, whatever and wherever that is - in the universe. They're all identically the same everywhere with no mistakes. We shall look into where they all came from in a little while.

This perfectly uniform identity of each atom of each element is just as well really. Every one of the atoms we're made of, as I write and as you read, will be totally replaced in our bodies as part of a natural wastage or wearing out process

CHAPTER 12 - THE KINGDOM OF GOD IS WITHIN US

we go through - within a year, on average. They're all lost and replaced by other identical atoms, each of their kind, each in the very same place in our bodies (give or take...) as the lost one came from. The cells of the stomach lining are all replaced every few days, bone cells every few years. All the while, I remain me and you remain you. If this were not so then we wouldn't last very long at all of course. In fact neither we nor any other living thing would ever have come into existence. And yet we did, we remain and - somehow we know it.

I will return to the question this raises of "Who then are we? Who is the 'I' of my existence whom I recognise as existentially 'Me' when all of me is constantly in flux? Where in fact am 'I' located in my body? Where is the "me" I know so well in my lifelong existence and even thereafter once having *"shuffled off this mortal coil"* as Shakespeare's Hamlet put it. Where in my body is my 'I'? It's like asking where in the universe is God? Yes, I will come back to this.

Each one of these 92 different 'elements' has its own unique combination of sub-atomic particles, of 'protons', 'neutrons' and 'electrons'.

Well, fine. But why draw attention to this? Might we not expect, in all the countless numbers of atoms and unimaginable distances spanning the cosmos, there might be some differences? Shouldn't there be some variation, even some mistakes along the way? Might not some 'protons' be different from others depending on where they are? Might not atoms of Oxygen or of Sodium for example, formed in different supernovae billions of years ago, billions of light years from here, be different depending on where they were made? No. We know this scientifically, that all the atoms of every element wherever they may be found are identical. There are no faulty 'protons', no faulty 'atoms' or any other fundamental particles. It's as if they were either made to be

Periodic Table of the Elements

Periodic Table

like this by some grand design, or they all just happened to somehow pop up in this way.

The 'periodic' table, cataloguing the elements in ascending weight order, was devised in the mid-19th century to map all the relationships between the 92 naturally occurring elements. Each element is given an atomic number depending on the number of 'protons' there are in the 'nucleus'. The first element is Hydrogen with one 'proton', followed by Helium with two 'protons' followed by Lithium with three on up to the rare, heavy but naturally occurring Uranium with 92.

There are some highly unstable man-made elements beginning with Neptunium with 93 'protons' up to Nobelium with 102. They remain in existence for no longer than a few hours, minutes or less without breaking down or decaying into lighter 'elements.' Their systematic structural continuity and ultimate natural limit, as catalogued in the 'periodic table' further confirms the extraordinary coherence of their design.

CHAPTER 12 - THE KINGDOM OF GOD IS WITHIN US

One may imagine there could be other life forms based not on the almost infinite number of organic combinations of Carbon, Oxygen and Hydrogen we have on earth, but on other elements *out there*. They could be made from other elements altogether, or even the same elements in a different form. But basically, these elements we know of are all there is - anywhere. Furthermore, we can say with some confidence we know enough about organic chemistry by now to rule out alternatives to the billions of Carbon-based molecules which constitute Life. It took close to 11 billion years after the 'Big Bang' to happen, but the eventual creation and evolution of Carbon-based life, derived from primordial Hydrogen and a little Helium, has been possible from the very beginning, 13.8 billion years ago. Some have even said, *"It's almost as if the universe has been waiting for us to happen."* We'll unpack this as we go.

Of course, the frontiers of science are always being extended and we don't know what we don't know. But certain scientific facts such as these have been confirmed over and over again over the decades and even centuries until they have been regarded in reality as being true. If, here on earth, an Oxygen atom were not made of 8 'protons,' 10 'neutrons' and 8 'electrons' it would simply not be Oxygen. One 'proton' more and it would just be Florine; not a very nice alternative to Oxygen, one 'proton' less and it would be Nitrogen. Each 'element' is in a perfect mathematical sequence with the 92 other natural, universally occurring 'ordinary matter' (as in all across the universe - although admittedly not at all the 'dark matter') elements. Scientifically speaking there never are any mistakes. They're all always absolutely perfect, in a sense kind of Holy?

Admittedly, this is not something we think about much, if at all. However, what I find quite intriguing is we are made from exactly what the earth, the moon, the stars and the galaxies are made of, as are all other living things down to microbes and even viruses. Not only this, but what we are made of was itself actually created in the white-hot (billions of degrees) inner parts of exploding 'supernovae' billions of light years away from here, billions of years ago, well before the earth began. And yet, Bingo! Here we all are! How? One might well wonder.

But how can we really know these atoms are all the same when they're so far away? Great question. The science of 'light spectroscopy' identifies every element by its unique signature, a rainbow spectrum of light. After decades of astronomical observations, every element has the same unique light signature wherever it is found, here on earth, and everywhere else in the visible universe.

Life on earth *as we know it* is composed of millions of chemical compounds all of which are based on the biochemistry of six elements - Carbon, Hydrogen, Nitrogen, Oxygen, Calcium and Phosphorus and a few others in combination. They make up almost 99% of the mass of all living cells. These hydrocarbons are all there is - to Life.

One may imagine there could be other life forms based not on this almost infinite number of combinations of .these elements. But basically the elements we know of are all there is - anywhere. *Other lifeforms* may make good science fiction, but even scientists accept they really are, fictional.

How then can this be? It is staggering. It's just incomprehensible to our minds. And yet - here we are. We can reasonably well comprehend how, if not necessarily why. As Einstein put it, *"The most incomprehensible thing about the universe is that it is comprehensible."* We know that we know. Furthermore, we know that we know that we know - as distinct from all the rest of creation. The Universe does not see us, but WE see the Universe. How can this be so?

CHAPTER 12 – THE KINGDOM OF GOD IS WITHIN US

Of Biblical Fact

"Is there anyone else up there!?"
Anon

Contrary to some expectations I do believe scientific fact, and our knowledge of the world we live in, is totally consistent with Biblical fact and our knowledge of God. This is what I want to explore in the rest of the chapter.

Speaking for myself amongst others, but by no means everyone, I believe science even supports faith at times when we may question our own subjective faith *reality,* our own *"cognitive bias"* in terms of *"Why did God allow……?"* or *"How long, Oh Lord!?"* and *"Is there anyone else up there?"* Doubt is not necessarily a bad thing in itself. *"I believe; help me with my unbelief!"* (Mark 9:24) is the place in which faith is nurtured through experiences of all kinds, not despite them, as I've recounted earlier. One may even cry out, *"Where was God when I really needed him?"* In the words of the famous *Footsteps in the Sand* poem, *"It was then he carried me."* Blessed is the one who knows this at the time. [41]

But I do see science as a kind of ultimate *"back-stop"* to doubt, to securing faith in a way which goes beyond our own subjective faith experience. Scientific principles by no means contradict the Bible. There's no real place to hide from God through Science. This is not to say some cosmologists, Nobel Laureates included, don't try pretty hard, although their confusion between God and Religion (man's interpretation of God) can be - well, confusing.

'String Theory' proposes not only an infinite universe but an infinite number of parallel infinite universes or 'multiverses.' Cycling into and out of existence in 10 or even 11 dimensions.[42]

Each works perfectly well with slightly different laws and constants from ours. E.g. the relationship between the radius of a circle and its circumference may have a different value for π. There may be other Me's or You's in any one or more of them existing simultaneously in these 10 or 11 di-mensions, not just our 4 including time. Either way this doesn't answer the big question, the elephant in the room, *"How did they all get here then?!"*

I'm reminded of the light-hearted story about the team of science professors and researchers and their post-graduate bearers. They're aiming to make a break for the summit of the highest peak in the highest of mountain ranges of knowledge. Standing on the shoulders of giants, they assemble their equipment, ropes, oxygen and support materials.

At the peak, they expect to find the final answers to the greatest questions mankind has ever pondered - the origin of the universe and the origin of life. They'd struggled, and they'd strived to overcome enormous difficulties. They'd never yielded to the temptation to give up, but were driven forward by the primordial human desire to overcome ignorance with scientific truth.

Any question about the meaning of life itself is not their concern now. The *"Why?"* of the assault on the summit itself is unnecessary. It's self-evident, *"Because it's there!"* The, *"What's it all for?"* somehow gets left behind as being unscientific - at least since Galileo proved the church to be wrong about the sun rotating around the earth. Some scientists are still quite proud to recount the story, and not without justification.

So the team finally achieves the summit, breathing what feels like their final gasps only to find instead of snow fields - green fields. There are lions and there are lambs lying down together in the grass as in Isaiah 11:6. They too are listening raptly, amongst the crowds of men, women and children of all races and creeds. They're all listening attentively to the calm, loving, powerful Jesus - the creator of all things the

CHAPTER 12 - THE KINGDOM OF GOD IS WITHIN US

"God of Love" (2 Corinthians 13:11), "the Spirit of Truth" (John 16:13). Thank you again BibleGateway.

The scientists and their bearers quietly lay down their paraphernalia and equipment in resignation, all their instruments, their Himalayan-spec laptops and digital texts and AI-driven algorithms. They sit themselves down amongst the crowd, listening intently. As their bone deep fatigue lifts from them, they hear what they'd always kind of known but could never quite admit to themselves or amongst themselves - especially since it might well mean going to church, worse still admitting their failures even to One who knows them anyway.

Jesus explains to them, "God is light" quoting 1 John 1:5 - God is not only the designer and the creator but also the very substance of matter itself, what quantum physicists refer to as 'gluons.'[43] Travelling at lightspeed, they literally glue together each 'proton' and 'neutron' in the entire mass of the universe, in our bodies and minds, our thoughts and even our memories. They are timeless particles since for an object travelling at light speed time does not even exist. 'Gluons', like 'photons', are literally everywhere all the time. They just ARE.

Jesus quotes from his word, written into the Bible in 1 John 4:8, *"God is love."* He explains to them what came earlier, that God's love IS the 'strong force' because *"in Him all things hold together"* (Colossians 1:17) all the 'protons' in all the atoms in the universe - by the very power of His love.

Meanwhile, they already know 'dark matter' and 'dark energy' are the other 95% of the universe about which as scientists they know nothing apart from their existence. Not that this other 95% is out there somewhere in some far-flung expanse of the cosmos beyond our own 5% corner of it. No. 'Dark matter' and 'dark energy' are here and now, all about us. All this is explained to them so they can understand

and participate. The Kingdom of God is truly *"within us"* (Luke 17:21 NKJV)

Before the crowd disperses to tend to their respective galaxies, a young boy showed up with five loaves and two fish, (John 6:9-13) and everyone ate their fill. (Well, I thought, why not?) The Master beckons the scientists over for a personal briefing full of praise for their tenacity, their selfless dedication to the search for and their love for Truth above all things. They are men and women after His own heart, made in his image, as they can see.

Much as they want to, He does not allow them to kneel. They stand in His presence as He introduces them to some of their heroes of *the faith* who'd gone on ahead, Galileo, Newton, Einstein, Bohr, Heisenberg, Hubble amongst many others, Archimedes, Aristotle, Hippocrates and there's Richard Dawkins skipping with joy and singing Hallelujahs, *"Now I see it all! Jesus is King of kings!"*

Steve Jobs, soon to welcome them there, had said just, *"Wow!"* over and over as he left his family and as he saw Jesus Himself reach out to him.[44]

Then again there was Dudley Moore, *"I hear the music all around me."*[45]

Jesus had promised all this to the Father who *"so loved the* (whole) *world."* He'd sent him to save all those he'd been given - to die for "once for ALL" - meaning once only for everyone. Much as they relished the thoughts of the Nobel prizes they would receive on their return, the scientists were reluctant to leave the Elysian fields. But eventually they departed, back to base camp - personally inspired by the embodiment of the Meaning of Life. One day, they would all come back to live there for eternity. They would explore the reaches of the known, and yet-to-be-known universe, the What, Which came before, the How and Who - and also the Why of He who'd created it.

CHAPTER 12 - THE KINGDOM OF GOD IS WITHIN US

Living Today in the Seventh "Yom"

Just to clear up one of many apparent contradictions in the Bible before moving on. In Genesis 1 the Bible famously declares Creation was all done in six *days*. The word translated from the Hebrew *'yom'* as *'day'* means also *'an undefined, indeterminate period of time'*. A similar usage in English might be *'in my father's day they used to...'* does not mean my father only lived for a day, of course. So, there were various phases of creation which can otherwise be described as epochs. Human ingenuity has devised many more than six such periods of time, but either way, "*In the beginning God......created.... the heavens and the earth*" (Genesis 1:1) as in everything in them.

In other words, "*This is what the Lord says. 'He who created the heavens, he is God; he who fashioned and made the earth, he founded it; he did not create it to be empty but formed it to be inhabited. It is I who made the earth and created mankind on it. My own hands stretched out the heavens; I marshalled their starry hosts.*" (Isaiah 45:18 and 12)

I make frequent reference to the 'Big Bang' of creation which happened 13.8 billion years ago. All the energy which became matter was released from a single point or "*singularity*" of zero size. There was no space within which anything could have any size. Then, within less than one quadrillionth of a second a full-sized universe came into being with a radius of 13.8 billion light years. Which is why we're just receiving light from almost its furthest reaches almost 13.8 billion years later. That was one BIG Bang! Then, 9.8 billion years after that there followed the 'Big Birth' the beginning of Life and 4 billion years of evolution - until now.

How did God do it? "*He breathed words and worlds were birthed. 'Let there be,' and there it was— springing forth the moment he spoke, no sooner said than done!*" (Psalms 33:9 TPT)

He wisely chose to describe the whole of creation in less than eight-hundred words in Genesis 1 for all people for all time in a way which is translatable into just about any language. Regardless of time or place in geo-history for any reader of almost any age. Regardless of their socio-economic or educational achievement or epoch of scientific knowledge in which they lived, the content of Genesis 1 is both comprehensible and significant in many ways. Most certainly it is not intended to describe 'Big Bang' quantum physics, nor what happened before the 10^{-43} seconds of the 'Planck Epoch' (named after Max Planck) before which the laws of physics didn't even existed.[46] There is proof scientific which says we cannot approach any closer than this to the beginning of time, let alone before such a *point in time* when there was no time. Nor does Genesis 1 provide a detailed explanation of the 'Cosmological Microwave Background' or CMB, the light echo of the Big Bang, to which we will return.

Nevertheless, actually believing what's written there in Genesis 1, having faith in its truth under all the circumstances of life, surely challenges the intellect. It challenges the heart, as it challenges the soul too, in its search for meaning.

I love the story of the man who'd literally reached *the end of his tether* in life, suspended over an abyss. The rope he desperately clung to was all that held him from plunging into the engulfing darkness. As his grip began to fail, his hands slipped inexorably down the rope until he came to a big knot at the end of the rope. His legs swung beneath him in empty space. At such moments, and sometimes not before, the inclination is at last to cry out in despair. Clinging desperately to the knot, and probably not for the first time he cried out, "*Someone! Help me! If there's anyone up there please help me!!!*" Time was running out very very fast. The gulf was about to swallow him whole if nobody showed up really soon. Then, in extremis of despair he cried out, "*If there is a God, please please help me!*"

CHAPTER 12 - THE KINGDOM OF GOD IS WITHIN US

The Palms of His Hands

Immediately, there came a deep, celestial, Fatherly voice from above. It advised him ever so gently but ever so firmly, *"Just let go, my child. I'll catch you."* To which, of course the man tremulously hesitated before making a final desperate plea, *"Is there anyone else up there?"*

Of course, there's no need to put scientific fact above Biblical fact as a substitute for faith. The Bible is either all right - or it's all wrong. The Bible explains itself as nothing else can. Faith is not a leap into the dark, rather a leap into the light. But, we live in life-challenging times. Disappointment can lead to doubt and to unbelief. Might we need to know from time to time if there's something beyond our own subjective *faith reality?*

The more we grow in scientific knowledge, the more impressed we can become with our own ability to know, instead of with the wonder of what it is we know. The less reliant on God we therefore risk becoming. However, I'm here to suggest the contrary. The more we know of God by studying, *the work of his hands,* the better we come to know who I AM (Exodus 3:14) says he is and who He says we are.

There is someone *up there* who wants to save us and can save us as in, *"The heavens declare the glory of God; the skies proclaim the work of his hands."* (Psalm 19 :1) We live in the seventh '*yom*' of existence.

The point of this chapter is to delve more deeply into the wonders of God's creation. Vast mysteries remain beyond both Biblical-fact and scientific-fact. Maybe they have not been hidden from us but rather hidden for us to find?

The saying, *"Man is the measure of all things"* is attributed to Protagoras in the time of Plato 2,500 years ago. It has been said, *"While there is nothing more certain than our very existence, there is nothing of greater significance in our lives as human beings than our own self-awareness."* This implies there can be no objective knowledge, there is no certainty apart from what is in our own minds. All knowledge is subjective, as observed by the observer.

Descartes' search for objective truth actually accentuated this conclusion and embedded it in our Western materialist culture. Deep insecurities in Descartes' own life put him on a quest for something so self-evidently true it could not be challenged. He needed something definite in life as a fixed and firm starting point in turbulent 1640s France.[47] His solution to this conundrum, the only thing he could really know for sure was, *"I think. Therefore, I am"*. Putting self as the only reality this consciously or unconsciously extends to what has become a modern mindset, *"Man/I - really am the measure of all things."*

CHAPTER 12 - THE KINGDOM OF GOD IS WITHIN US

However, we don't need to look up at the night sky for very long, nor even beyond a mirror, to see the fallacy there. Even so, one may say, *"Why do we need a God anyway?"* Well, if God is God, the creator of all things, his existence cannot be dependent on our need or lack of need for him. He was, and is, and is to come - or not. What's so is so, quite independently of our own needs and unconscious bias towards *masters of the universe* thinking. And yet the *need factor* can become a determinant of truth in few settings other than this.

Fortunately, as created ones, I believe God does know our needs. He came here in person as Jesus, literally *"in the flesh"* not just to see for himself, not just so we could know he knows, but to do what it took to be with us always - through it all. Might the man hanging above the abyss be inspired then, as I am sometimes, to heed the reply he heard, *"No. I am all there is. But my Grace is enough. Let go. I will catch you."*

As we know, there is vast emptiness between the stars and galaxies in the external visible universe. Similarly, there is vast emptiness in and between the atoms in the internal invisible universe of our own bodies, as with everything else around us. Yet, we find a staggering orderliness and mathematical consistency across them all which totally defies chance. Who can this Creator be?

I will propose all and everything in existence is contained within His Kingdom. The Kingdom of God *"in which we live and move and have our being"* (Acts 17:28) is also *"within us"* (Luke 17:21 NKJV) just as are the 'dark matter' and 'dark energy' which also permeate the universe.

However, I'm not looking to convince anyone and get to the *right answer* about how all this relates to a Kingdom of God. I plan rather simply to share with you a part of my own journey of enduring faith.

Occasionally in our lives this context can of course become a deeply significant challenge to faith, on behalf of ourselves and of others. I'm looking to share how I navigated this journey to the very backstop of faith by matching scientific facts together with Biblical facts. I believe science simply and quite accurately describes the visible part of creation which is itself merely a small part of the mostly invisible Kingdom of God, and our physical place in it. If God really does exist how could science and the Bible ever be at odds with each other since he created them both - a solipsistic argument not offered as proof but simply for illustration.

Scientists freely admit there are very many questions yet to be answered, of course. In fact, one of the key tenets of the scientific method is to constantly question and probe the boundaries of knowledge for inconsistencies. It's at the boundaries where discovery lies. Scientists also agree certain questions can never be answered in scientific terms. The "Whys?" of existence and the meaning of life. However, while science cannot provide proof of God's existence it does provide plausibility.

For example, and as I will detail later, Time and Space themselves did not exist before the 'Big Bang'. Since energy takes up no space at all they both began from what's called a 'singularity.' This is a point of zero size containing enough energy to be converted to the entire mass of the universe to fill, or rather to become, all the Space of the Universe. In that instant too Time began. Before this there was no time. Time and space just did not exist. What came before time, space and the laws of science began? Into what and from what is the universe expanding? Answers to such questions are defined by science to fall beyond the realm of science and therefore into exclusively philosophical, mathematical or spiritual realms. But they make really interesting reading.

CHAPTER 12 - THE KINGDOM OF GOD IS WITHIN US

There is actually considerable agreement amongst most astrophysicists. All the galaxies are moving away from us at accelerating speed. Ours, therefore, might seem to have a central location. However, every galaxy is accelerating away from every other galaxy, so they can't all be at the centre of a three-dimensional universe just as we are not ourselves.

Rather, it is proposed by way of illustration, the universe as 'spacetime' is the surface of a sphere - which is constantly expanding in diameter. Each galaxy is located on this spherical surface of 'spacetime'. As it continues to expand space itself, the universe expands. Even as the surface of this somewhat hypothetical sphere expands and grows it does so at an accelerating rate. The universe does not expand into anything at all since it all just exists already on the apparently endless expanding surface of an expanding sphere. But one is still bound to ask, what lies within the sphere, and all around it - but God? Where can the person of God be found in the universe? We might also ask where can *I* be found in my body? Either way, He exists as do we.

As science itself teaches us, such metaphysical realms cannot be detected by measuring instruments however powerful or sophisticated. Neither can they be defined in mathematical terms, especially as we draw in toward the meaning of our existence.

We are equipped to pick up the multitude of signals constantly emanating from our household or office browsers not by means of our physical senses but by our computers or smart devices. We can say likewise, we are designed and equipped to pick up the multitude of signals constantly emanating from God by spiritual not by scientific means since science has quite rightly vacated that space.

We are indeed designed and equipped both to know God and indeed to need him to fill Pascal's renowned '*God-shaped hole*' in each one of us. "*God created mankind in his own image, in the image of God he created them; male and female he created them."* (Genesis 1:27) So he did, from His own perfect, composite amalgam of male and female.

Here we might characterise God as the artist, Jesus as the model and ourselves as the image. Although we are like Him, He of course, is not like us. Nevertheless, we are each one of us His masterpiece.

In this Godly divine and human semi-divine context, Religion is man's multiform interpretation of God's intent. Although somewhat unkindly, Religion can be characterised for better and for worse in this artistic illustration, as the multiple responses of the art critics.

It's this *straw man* art critic of Religion, as expressed collectively in Christianity by the Church and individually by its members, which tends to fail under public scrutiny. Sometimes, there are spectacular failures of Religion in general and the Church in particular through human weakness. Given the context of creation we've been looking at it's hardly logical, but this human fallibility can invalidate (for many) the very existence of God. And this despite the profound distance of kind between the *artist* and the *art critic*.

This contrast between God, the artist, and the interpretations of him in religious circles are far more profound than can be found anywhere in the art world. Nevertheless, direct or indirect experience of misadventures in church life can seem to count for some as prima facie evidence against the very existence of God. The theory instead then becomes, "Everything around us just is, just as it is and just happened the way it is - either all by itself, or it's always been that way anyway."

In times of grief, this '*God shaped hole*' yearns more than ever to be filled. One purpose of this book is therefore to reset the context away from our sometimes-regrettable individual

CHAPTER 12 - THE KINGDOM OF GOD IS WITHIN US

experiences of church and the negative conclusions far too often drawn from it. Hence my apparent displacement therapy of looking to the jaw-dropping realities of our existence. It has both motivated and allowed me to at least begin to consider here the outer universe of 10^{23} stars, *"known by name with not one of them missing"* (Isaiah 40:26) and the 10^{-23} metre-wide flawless subatomic universe within. It led me on to consider what the implications are of twenty times more 'dark matter and dark energy' of which we know next to nothing - might be.

So, in the end does God really exist anyway? Again, I'm not looking to prove or disprove what's beyond the realm of science, what can neither be scientifically proved nor disproved. I think most of us would agree there's no logical connection at all between something being in existence and our desire for it to exist or a belief it does. London exists whether I've been there or not. London exists whether I just want to go there, or whether I believe it's there or not. We should therefore be able to make the proposition, if God exists he does so whether I believe it or not. If God does not exist, he does not, whether I believe so or need it to be so, or not. On the contrary, it almost goes without saying, if God does exist it is not a matter of opinion, nor of anyone's needs or personal experience, but simply as a matter of fact.

I'm looking, instead, to map my own 'reductio ab absurdum' approach to faith whereby, despite all my experiences of recent years since Charmaine's diagnosis, it becomes simply absurd for me to think God does not exist. Meanwhile, *"I know in part; then* (one day) *I shall know fully, even as I am fully known."* (1 Corinthians 13:12)

Let me just add once more, I believe such statements are not true simply because they're in the Bible. I believe they're in the Bible because they're true, put there by the One who is Truth.

Thought of in this way we know, however intuitively, it's all but impossible for the vastness, complexity and consistency of creation to just happen to be there all by itself, without a Creator. There's a compelling kind of logic in, '*Stuff like this can't just happen on its own.*' As the atheist reportedly said to the Christian after a long debate, "*I'd better go and make some tea. It won't make itself.*"

A some***thing*** as a Creator is somehow subjectively unsatisfactory. The human mind turns naturally to a *Some**one** must have made all this.* The word God springs, however tentatively, to mind. Maybe it has been dismissed many times, and this assumption does not make God so. But maybe it's time to reconsider?

We might well ask, "*Who are we to say a God who can create something so vast as the Universe and as complex as Life can know us, love us and share His thoughts with us?*" (Amos 4:13) Why would he?

Equally, "*Who are we to say a God who can create something so vast as the Universe and as complex as Life can NOT know us, love us and share His thoughts with us?* Why wouldn't he? "*He so loved the world....*" (John 3:16) These are not just rhetorical questions but ones we will explore further.

But even so, there are so many of us thinking so many thoughts all the time. How can God, even as big as he is, share them all even if he wanted to? Great question. I'm glad we got to it at last. The answer is in the word *time*. Having established God, as hopefully we have, as the creator of time His life does not consist of one moment following another, as it does for us who are within time. Every moment, from the beginning of time as He made it, is always present to Him. He has all eternity to listen to the split-second thoughts of everyone everywhere who ever was, or is, or ever will be. He therefore has an infinite attention span for each one of us -

individually - even now. He can therefore share his thoughts with as at the precise moment in our thought time-frame to connect with them and make sense. That's just the God we have!

C. S. Lewis illustrates this well in his book Beyond Personailty, *"Picture Time as a straight line along which we have to travel. Then picture God as the whole page on which the line is drawn. We have to move from point A to point B on the line and immediately on to point C in time. God, from above or outside or all around, contains the whole line and sees it all."*

But Where's the Evidence?

Some still ask, *"But how can we believe in God when there's no evidence?"* What evidence might we need further to what we've discussed so far, from the wonders of the quantum world to the intergalactic extensiveness of it, not to mention the otherwise absurdly long odds against our own chance existence? The kind of *copper-bottomed* certainty of the existence of God, based on the kind of laboratory evidence the questioner presumably has in mind, would eliminate all doubt - and all choice. It would also eliminate the freedom of choice between being right and being wrong which we as humans all esteem and aspire to and hold most dear.

Wouldn't such absolute certainty lead to a world of automatons held captive to undeniable truth? The tyranny of undeniable Scientific Truth would simply replace undeniable Religious Truth. The *"God who so loved the world"* (John 3:16) seeks our love in return. Yet it takes little imagination on our part to know you cannot force someone to love you just by over-powering them with your love. Happily enough instead, *"He is wooing you ......to a spacious place free from restriction to the comfort of your table laden with choice food."* (Job 36:16) I believe this *"choice food"* includes the delectations of the search for truth of the kind we '*know in our knower*' - if I could put it like this. It's often acknowledged as the very essence of a Nobel Prize.

The following will have nothing to do with religion as a human construct, of making God in our own image. It will have everything to do with God having made us in His image, with our place in His Kingdom, and with the place of His Kingdom in us. "*God SO loved* (all) *the world he gave his only son...*" (John 3:16) To quote the poet Jim Morrison, "*We're all in search of someone who already found us*" - or one who never even lost us. Granted, it can sometimes seem God has become distant. But then, if that's the case, guess who moved?

As Christians we read, "*I in them and you in me.*" (John 17:23) Jesus is in us, and God is in him. What on earth does this mean for us here and now? What is this profound mystery of how God and His Kingdom is not only at hand but actually within us? (Luke 17:21 NKJV) Later on I will describe even further the timeless particles called 'quarks' these ripples or eddies in a universal force field, the very stuff of which our 'protons' and 'neutrons' are made.[48] This will prepare us for the following chapter on "Jesus and our Glorious Bodies." There I re-emphasise God's love as being the 'strong force' holding together all the 'protons,' 'atoms' and 'molecules' in our bodies and all the matter in the universe since, "*God is love*" (1 John 4:8) and "*in him ALL things hold together.*" (Colossians 1:7)

Insofar as all of us are created by God, *the artist,* in *the image* of Jesus, *the model* I described, religions look to understand and respond in a variety of ways to a creation story. Not surprisingly, since God communicates with all creation, most religions therefore have much in common however well or imperfectly we listen to and comprehend them. However, only in Christianity does God Himself as Jesus choose to come here, to be simultaneously 100% God and 100% man, to live like one of us and die instead of us.

Again, as in the Joan Osborne song,[49] *"What if God was one of us? Just a slob like one of us. Just a stranger on the bus trying to make His way home?"* Indeed, he did become one of us. *"The Word (Jesus) became flesh and made his dwelling among us."* (John 1:14) He did this in order to be with us and to be like us in many but not every way (100% human and 100% God.) He came here to show us where is home and how to get there since He Himself IS *"the way, the truth and the life."* (John 14:6)

Jesus surrendered his very life, publicly flogged and nailed naked on a cross for three days to suffocate to death. Why? He did this to pay the ransom, to pay the price for all the wrongdoings we as individuals have had committed against us, and by us (1 Timothy 2:6) - out of Love. For my own part I believe Jesus is the Son of God and we in turn have somehow been given the right to become adopted as children of God. (John 1:12) As C.S. Lewis put it in Beyond Personality, *"The Son of God became a man to enable men to become sons of God."*

Jesus was arrested in the well-known Garden of Gethsemane. He chose not to escape. He could easily have disappeared into the endless open desert stretching out right there behind him, unhindered all the way to Baghdad or into the desert places of Saudi Arabia. It was all part of the plan from the very beginning. God had addressed the devil in the other equally well-known garden, the Garden of Eden. Speaking ultimately of Jesus He prophesied that, "He will crush your head, and you will strike his heel" (Genesis 3:15) Less well-known is that Jesus as God Himself(!) eventually fulfilled His own prophesy. Satan was defeated. His head was crushed at the cross just as Jesus' heels were nailed to it.

Even more fearfully Jesus chose to put himself into *"the great winepress of God's wrath."* (Revelation 14:19) He paid the price, God's anger at all we have ever done or will ever do to harm, or be harmed by, others.

This is why Jesus declares *"No one comes to the Father except through me."* (John 14:6) If there were any other way to the Father Jesus would not have let himself be crucified to death. In Gethsemane he even pleaded with God, *"Isn't there some other way? Take this cup* (of your wrath, over how they treat each other) *away from me."* (Mark 14:36) In prayer over having to face God's wrath (which he knew only too well how terrible it could be considering just what we do to each other) this we're told, *"his sweat was like drops of blood falling to the ground."* (Luke 22:44) But he put himself aside and stuck with the plan. It was - the only way. *"Yet not what I will, but what you will."* (Mark 14:36)

This way of the cross, through Jesus, is *the only way* for us to have eternal life. If there had been any other I'm quite sure God would have found it. I will come back to this point, to explore what I think this means in scientific terms.

Jesus shows us the way by saying, *"I AM the way and the truth and the life. No one comes to the Father except through me."* (John 14:6) So it seems a bit odd to want to go to all the trouble of finding one's own way to the Father. He, the very one we want to be with, tells us there is only one way to reach him. Why not welcome the one who sacrificed his life to become *"the way?"* Just asking.

It has been said, *"If you really want to go from Manchester to London then don't board a train going to Glasgow."* Both are cities, and have this in common, but admittedly they are very different places. In simple terms taking the train to London (Jesus) is the only way to get to London, (the Father) Unless of course we want to spend the rest of our days getting on and off different trains to other different destinations and starting again, so to speak. If, and only if, London is where we really want to go, why not just take the train which is going there?

Why would we want so much to go and be with someone we don't actually trust enough to tell us how. If Jesus is not the One who created everything why bother? If he really is the One then what hope do we have of finding a better way on our own? Well, there's another great question to ponder until we return to it later.

Some would say there are many different ways to God, and many Faiths describe different ways. But although there are many similarities between them, there are some quite contradictory differences too. They can of course all be wrong, but they cannot all be right.

Jesus did not go through all he did when all the kind-hearted people around the world who might well care for each other quite well without him. We do fail (miserably sometimes) to behave towards others as we know we should, even to our loved ones. Others fail too (miserably sometimes) to behave towards us as they should, even our loved ones. We really struggle sometimes. We struggle most of all just to forgive them for their shortcomings against us. Most of all we can struggle to forgive ourselves for our own shortcomings towards them - and even towards ourselves.

Jesus put it like this, *"In everything do to others as you would have them do to you, for this sums up the Law and the Prophets."* (Matthew 7:13) As in *that's all*. That's it, everything, all he asks us to do. In other words, *"Love one another. As I have loved you, love one another."* (John 13:34) I think it's fair to say we need all the help we can get to do this, every day. Even *"Love your neighbour as yourself"* (Matthew 19:19) is challenging. How well do we really love ourselves and how well can we do it without his help?

Knowing the father-heart of God, His character, we know He is righteous and just. His *"righteousness is like the highest mountains, his justice like the great deep."* (Psalm 36:6) As human beings we, like him, have a powerful sense of justice and especially of injustice. From a very early age we know when we have become the victims of injustice. Amongst the first few words we learn as a child are, *"That's not fair!"* Our wrong-doings against each other have consequences, sometimes most severe ones. A moment's thought will illustrate from everyone's experience just how serious they can be. We too demand justice be done to those who are unjust towards ourselves and others. We have a strong natural sense, wrong-doers must be punished as part of putting wrong to right.

But God is not only *"righteous and just."* (Psalm 50:6) *"God is Love"* (John 3:16) too. God not only loves, He IS love. How then can harsh Justice be reconciled with Love?

What is this Thing Called Love?

> *"Love is large and incredibly patient. Love is gentle and consistently kind to all. It refuses to be jealous when blessing comes to someone else. Love does not brag about one's achievements nor inflate its own importance. Love does not traffic in shame and disrespect, nor selfishly seek its own honour. Love is not easily irritated or quick to take offense. Love joyfully celebrates honesty and finds no delight in what is wrong. Love is a safe place of shelter, for it never stops believing the best for others. Love never takes failure as defeat, for it never gives up."*
> (1 Corinthians 13:4-7 TPT)

God can be and do no other than Love. The heart of love is forgiveness. However, since *"His justice is like the great deep"* (Psalm 36:6) God must therefore exercise justice and so must punish all the offences we do against each other. He is required therefore to both punish and forgive - at the same time.

So, God has this huge dilemma. He needs both to exercise justice while punishing the very ones he so loves and therefore wants to forgive. How to reconcile Justice with Love? Jesus substitutes himself for us. He substitutes himself for the harshness of God's punishment for our injustices, against each other and from each other. This is *the only way*.

Jesus, on the cross and through the cross, took upon himself the punishment for all our wrong-doings. We are now free, through his substitution, to forgive those we have been wronged by. In turn we can receive forgiveness from those we have wronged. The good news? We can do it through the power of God's love not our own, such as it may be. The mere act of saying to the unforgiven about the unforgivable, '*I forgive you*' whether we mean it in our hearts or not, can be enough to release that transformative love and forgiveness into the lives of both. '*Mere*' sincerity can and will follow later.

But, let us not forget. Jesus, "*the Messiah is God.*" (Romans 9:5) Having become himself manifest on earth as Jesus the Messiah, God Himself became the atonement for our failings and wrongdoings, for breaking His law - which is to "*Love one another as I have loved you*" for "*whoever loves others* (in this way) *has fulfilled* (all) *the law.*" (Romans 13:8) It's just that we can't do that, so He did.

I spoke earlier of time, the beginning of time at the 'Big Bang' of the 'Planck Epoch' of within 10^{-42} seconds of the 'Big Bang.' Looking back from the 'superiority' of our own viewpoint it is often said that, '*The laws of physics break down at that point.*' More accurately, within that 'Planck Epoch' looking forward from God's viewpoint, they had not even been created. This leaves us with the apparent mystery of what was going on before that, before time even began. Well, this reconciliation between Justice and Love, between punishment and forgiveness was planned with us in mind "*before the beginning of time.*" (2 Timothy 1:9) Yes, in there because it's true.

Jesus suffered God's judicial punishment in our place in order to put right all our wrongs, for us to be forgiven not punished. Judicial, that justice be done - in love. In human terms, this is a contradiction every parent understands when raising their children. In terms of our eternity past and eternity future, it's a problem Jesus alone could resolve for us.

He paid the ransom to redeem us, to free us from the consequences of wrong-doings, both as victims and perpetrators. One life sacrificed for many and even so a life for a life. His life for my life.

What remains for us to do is to reclaim our ransom, like reclaiming a lottery ticket. But in this instance every ticket is free, and every ticket is a winner! They're free to us and ours to claim, because Jesus already paid top price for each one of those tickets - individually.

Hopefully this goes a little way further towards explaining the controversial exclusivity of Christianity and of Jesus being *"the (only) way, the truth and the life. No one comes to the Father except through me."* (John 14:6) Stipulating anyone who wants to come to the Father must go through Jesus seems *non-inclusive* and to *discriminate* against those who want to make their own way to the Father by a different route. Put simply, it's almost like wanting to play rugby by joining a football club and complaining about their rules and the shape of the ball. Either way everyone's welcome to a seat in the stadium. Why insist on paying for *"life and life to the (very) full"* (John 10:10) when it's there for free, since Jesus paid such a high price for us to have it? But we do have this way of wanting to.

And so, to the Trinity. Although this is not a word to be found in the Bible. Jesus is God. (Romans 9:5 NKJV) As, *"In the beginning was the Word, and the Word was with God, and the Word was God."* (John 1:1)

CHAPTER 12 - THE KINGDOM OF GOD IS WITHIN US

I understand the person of the Holy Spirit to be the very intensity of the relationship between God the Father and God Son made manifest as a person. Furthermore, our bodies are no less than temples of this very same Holy Spirit. (1 Corinthians 6:19) "*He is in you, whom you have received from God.*" (1 Corinthians 6:19-20) Yes, as it says. So, I make frequent reference to the Holy Spirit.

This is the same Holy Spirit who "*Hovered over the waters*" in Genesis 1:2 when "*the earth* (universe) *was formless and empty, darkness was over the surface of the deep.*" Or as astrophysicists put it, a dense plasma cloud of 'quarks' and free 'electrons' blocked out all light, and darkness filled the universe, as I will explain further in a moment.

But just to be clear about the identity of Jesus, "*The Son is the image of the invisible God, the firstborn over all creation. For in him all things were created: things in heaven and on earth, visible and invisible......all things have been created through him and for him.* "*He is before all things* (before even time itself) *and in him all things hold together*" (Colossians 1:15-18) *(referred to earlier as being the 'strong force.'* Hence, through these apparently contradictory scriptures, the perfect identity between God (who said....) the Holy Spirit (who hovered....) and Jesus (through whom and by whom all that was made was made - the One who died on the cross) is revealed as One.

The 'Big Bang' created temperatures of 10^{32}°C immediately after the 'Planck Epoch' I just mentioned, 10^{-45} seconds after it happened. The whole universe was filled with a dense fog, opaque clouds of 'plasma' of free, primordial, pre-atomic 'quarks' and 'electrons.' For the first 380,000 years this blocked out all the light from the original 'bang.' There was darkness over the depths of space, or "*Darkness was over the surface of the deep.*" (Genesis 1:2)

Light could not penetrate this plasma fog until, in the so called 'recombination epoch' the temperature of the Universe dropped below 100,000,000 degrees. Every single 'proton' in the universe combined with two 'electrons' to form individual atoms of Hydrogen. A single 'photon' of light was actually created and emitted every time this happened - for as many protons as there are in the universe. We know this because of course we can quite easily simulate such events. Something similar happens when Hydrogen is converted to Helium. The fusion process fuels the stars and our own sun, the very light which drives photosynthesis in plants and keeps us all alive. Just then, like raindrops condense from and disperse a rain cloud, these newly created Hydrogen atoms condensed out from the plasma cloud. This dispersed the plasma which had blacked out the original burst of light from the 'Big Bang' and created still more light in the process. The Universe became transparent to its own primordial light simultaneously creating even more new light. Just 380,000 years after the cataclysmic beginning there was light.

Before then everything, *"was formless and empty, darkness was over the surface of the deep, and the Spirit of God was hovering over the waters. And God said, 'Let there be light,' and there was light. God saw that the light was good..."* and it went on from there. (Genesis 1:1-3)

As mentioned at the outset the Bible was written at least 4,000 years ago. It had to be in an accessible, allegorical form for pre-scientific generations of readers all around the world. Still relevant today I believe God is simply saying to us, "Don't bother your head too much about all the details for now. I was already there, before the beginning of time. Then I just said so, and time and everything else began." Instead, I think God wants us to grasp the really important themes of Genesis 1 and 2, such as love, forgiveness, meaning in life and Jesus, for our everyday lives - and beyond. A timeline is

CHAPTER 12 – THE KINGDOM OF GOD IS WITHIN US

shown here to catch up on the kind of detail found missing from the Bible.

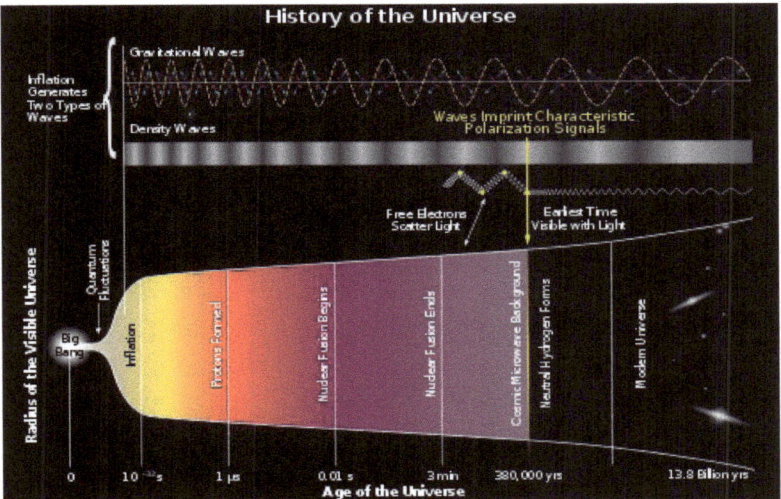

History of the Universe

So, as illustrated here and throughout, I propose scientific truth to be consistent with Biblical truth. Things are not true just because they're in the Bible. Yes, that's right. They're in the Bible because they're true. Although still full of mystery, the Bible reveals and describes these external and internal universes. It has been said, "When all else fails, read the operators' manual." As *the operators' manual for life* the Bible takes us quite beyond the realm of science into meaning. Science, as far as it goes, remains in its turn, true to the Bible.

As you will hopefully see when you read on, I'll quote a few more scientific facts and refer a bit more to Einstein, which alone can make our eyes glaze over as in, *"I never was any good at science at school. I never did understand any of this stuff."* But let me try to reassure you, no unusual scientific knowledge is required. Hopefully it will be enough to simply hold your nose and follow step by step.

So, *"Where are we? Where did all this come from anyway? Who are we? Where are we headed?"* For myself this is especially relevant as one of the *"ones left behind."* My wife went on ahead and I at some point am bound to follow. Of course I'm curious to know where.

Where Are You Now?

> *"Nowhere" begins and ends here....*
> (Anon)

As I mentioned before, although it doesn't seem like it, we do live precariously on the mere surface sheen of this little droplet of molten magma speeding along around the sun and the sun around the Milky Way at the combined speed we calculated earlier at nearly 700 times the speed of sound. The earth is 150,000,000 km from the sun but barely 13,000 km across. It can be compared on this scale to a piece of grit in your shoe. However, you probably wouldn't even notice a piece of grit quite this small. Although the lives we live and the world we live in are so abundantly real and important to us, in the scale of the created universe - we are of course far from the main thing we seem to ourselves to be.

This 15km to 20km thick surface sheen on this droplet of magma is what we call the earth's crust. It's barely 0.2% of its 13,000km diameter. It wouldn't even rival the skin on the proverbial rice pudding. What seems to us like solid ground compares less with the thickness of an eggshell, and more with the thickness of the membrane coating the inside of an eggshell. Tiny wrinkles appear on its surface from place to place. So much bigger than we are, we call them mountains. But in reality, forest couched, snow-capped, high and ever so mighty as they may seem, they are but the crispiest part of the earth's so-called crust. If we ever see as much as say the top 6,000 ft. of a mountain it seems enormous. But even Mount Everest's full 29,031 ft. above sea level at barely 0.07% of the diameter of the earth, is merely afloat on molten magma.

CHAPTER 12 - THE KINGDOM OF GOD IS WITHIN US

10 km (32,800 ft.) is a fairly short walk, so let's go for a walk, but not on the earth's surface. Instead, let's head vertically down into the so-called crust, this fragile, tenuous membrane. Here we discover a place where the temperature exceeds 250°C. Taking instead the same relatively short walk 10 km vertically upwards, the temperature drops below -65°C. Most life as we know it would end long before reaching those places just 10km away from where each of us are safely sitting even now.

Nevertheless, as emphasised by findings on climate change, the average temperature on the surface of our tiny globule we call earth has not fluctuated by more than 1°C to 1.5°C in thousands of years. Temperatures range nowadays between +50°C to -50°C. However, this 1°C average variation is less than anyone could detect if the temperature in the room changed by quite so little. Fortunately for us, until humanity recently interfered, it remained set at a *Goldilocks* temperature, neither too hot nor too cold but just right - for life.

As the earth's interior cools very slowly and shrinks, its core very gradually crystallises. It stirs up the magma's molten iron as a 'geodynamo.' It generates powerful electric currents which create a magnetic field or 'magnetosphere' enveloping the whole earth. It stretches far out into space and well beyond the visible Aurora Borealis part. Fortunately for us, it stops the solar wind of charged particles from the sun from stripping away the earth's atmosphere until, like Mars there'd be none left.

Good for us, this precious atmosphere contains the Oxygen we breath to stay alive, it having been breathed out by the plant life which breathed it in from us. In a complex diurnal chemical reaction the Chlorophyl in plants and trees nicely converts the waste Carbon Dioxide they breathe in from us during the night back out as Oxygen during the day for us to keep breathing it back in - all the time.

Enzymes such as Chlorophyllase accelerate parts of this and similar life sustaining processes to milliseconds which would otherwise take millions of years.[50]

We might well continue to wonder, *"How did all this happen? How come it's all so perfectly 'designed?'* so we and all of life around us, can live and thrive in a nowhere solar system in a nowhere galaxy amongst thousands of billions (not thousands and billions) of others? How on earth did we ever get here? Nowhere else in our solar system do perfect conditions of just the right, stable temperature, renewable oxygenated atmosphere and so many other conditions exist in perfect self-sustaining equilibrium. There are no seas of Ammonia or rivers of liquid Helium here as on other planets. How ready might our spirit selves be to take our chances anywhere else even in our galaxy, or in any of the others in this mere 5% fraction of the universe we know so much about - once body and soul have ceased to be?

The more extreme the complexity and interconnected our natural world and life on it is seen to be, the more absurd it can seem that God somehow arranged all this over all those billions of years - just for us to be and to see. And yet, the more extreme the complexity, the more we have to wonder where did it all come from anyway?

Chapter 13

Where did all this come from anyway?

The 'Big Bang' is still 'banging.'
Michael Morris

Well, I'm glad you asked really. Going back again to the beginning. The scientific community broadly accepts there was a 'Big (as in very very big indeed) Bang', about 13.8 billion years ago.

"Cosmic inflation theory" has it that *"In the beginning"* the entire mass of the universe, of all 'spacetime' itself, became totally present - everywhere. From not being there anywhere it became there everywhere. It went from a 'singularity' of zero size and enough energy to being the entire mass and extent of the universe.[51]

How can this be? How can the mass of the entire universe - which is a lot - be concentrated in a point of zero size? It would need to have infinite density. However, energy has no size. Present in this 'singularity' became all the energy ever needed to create the whole mass of the universe. At

Big Bang

the 'Big Bang' moment of Creation it was ALL converted to mass or matter from this primordial energy. We know from Einstein's famous equation $E=mc^2$ mass and energy are interchangeable. The m is mass and c^2 the speed of light,

(186,000 miles/second) multiplied by itself. That's an awefull lot of energy E, especially when it's five times more than the mass of what we can see of the universe using even the latest James Webb Telescope.

So, there was this singularity; finite energy however huge, but zero size, zero mass and therefore zero gravity to oppose what suddenly happened next - this humungous 'Big Bang!' Gravity acts ever more strongly on pieces of matter the shorter the distance between them. At a near zero distance, the gravity created by that amount of matter should have meant it could never happen. However, it evidently did. In the beginning, when *"God said...."* matter was created even faster than gravity could have stopped it ever happening, even before gravity and the other natural laws came into being.

CHAPTER 13 - WHERE DID ALL THIS COME FROM ANYWAY?

The Bible as we know describes it like this in non-scientific terms, "*In the beginning God created......*" time, matter (both light and so-called dark) and even space itself - from His own energy. Might we not add He continues to create 'dark energy' - from His own energy? Well, if having accepted the foregoing, from where else?

So, time just began. All this energy suddenly converted (itself?) into matter in proportion to the speed of light multiplied by itself. But it happened very much faster than the speed of light. Light, travelling through space is limited to no less and no more than 186,000 miles per second. However, space itself, and all the universe within it (!) suddenly became 13.8 billion light years in radius, 27.6 billion light years across. Hmm, I hear even myself say.

However, according to 'cosmological inflation theory' this happened in the 10^{-36} seconds, between 10^{-34} and 10^{-32} seconds after the beginning of time, not before then and not after. As a reminder, the minus sign before the little 36 means 10^{-36} is the same as 0.1 with 36 zeroes before the 1. So, this 'inflation' to 27.6 billion light years across happened in just 1/1,000,000,000,000,000,000,000,000,000,000,000,000th of a second.

There seems to be a problem here, because Einstein's 'Special Relativity' also tells us it's impossible for anything to travel through space faster than the speed of light in a vacuum. But we have to remember, space itself through which light was to travel, expanded faster even than this. Space itself was created, came into being, faster than light could travel across it, faster than gravity was created to hold it back. This totally amazing and counter-intuitive idea referred to in typically prosaic terms as Inflation Theory won for its creators (Starobinsky, Guth and Linde) the Norwegian equivalent to a Nobel Prize, the Kavil Prize in 2014.

The universe has continued to expand over the intervening 13.8 billion years since then, although at a very much slower rate. It's speed of expansion is nevertheless being steadily accelerated by the relentless expansionist force of 'dark energy'. Light coming from the most distant galaxy is being observed just now by the recently launched James Webb telescope. Referred to as JADES-GS-z13-0 it's 13.6 billion light years away, which means it's been 13.6 billion years since the light we see actually left there.[52]

We're seeing it now from here as and where it was 13.6 billion years ago in the past. From having a radius then of 13.8 billion light years and therefore a diameter of 27.6 billion light years across the universe has continued to expand by a further 33.6 light years, albeit at a slower although accelerating rate.

This puts the current radius of the visible universe at least 13.8+33.6=47.4 or 94.8 billion light years across from side to side, something we might struggle so understand from press reports.[53] Well, I thought I'd just throw this in since these numbers are being bandied around quite a bit in the popular press nowadays without always being fully explained. And anyway, it does rather put things and God into perspective......

Again to paraphrase two earlier questions, "Who are we to say God can and is......?" versus "Who are we to say God can NOT and is NOT...?"

So, neither matter, nor energy, not even time ever existed before the 'Big Bang.' As we know, science has tracked events down to 10^{-45} seconds known as the 'Planck Epoch'.[54] But there was nothing as in no**thing** at all before it. Afterwards, yes. But there is no way to know anything before this - at all. Scientifically and logically speaking there was no time before time began. The laws of physics, by which to ever know what came before them, had not yet been created. The time before time began has no practical meaning in the physical world as we know it - apart from God, by any other name.

CHAPTER 13 - WHERE DID ALL THIS COME FROM ANYWAY?

So, there's little left to say about the myth of science that, '*God does not exist because we have evolution.*' The myth of Religion that '*Evolution does not exist because we have God*' should need less energy to dispel.

To summarise, these are weighty conclusions. Within the very first second of the 'Big Bang', but not before, all the matter of the universe, the laws of physics and all the forces and quantum energy fields, we'll go into later, had come into existence. All the energy needed to create all the mass of the entire universe - at once - had been expended. To be clear, (if one can be about such things) it was not just all the matter of the universe expanding simultaneously into space. Space itself was expanding from itself, as nowhere just became somewhere.

But we have a God who can, "*call into being things that were not as though they were*" (Romans 4:17 NKJV) and now they are.

What better way to describe "*things that were not*" than what science refers to as a 'singularity?' In other words, the Universe came from no**thing**, from something which has been scientifically proven cannot be known to science.

Yes, it makes my head hurt, too. It took me a little while to find out myself how the universe became 94.8 billion light-years across when it's no more than the 13.8 billions of years of age. Since one light year is the distance light travels in a year it should be no more than 13.8 billion light years across.

It's all part of what happened on just the first '*yom*' the first time period or the first day.

As the Bible puts it, "*In the beginning God - created....*" He was there already before the beginning before He created time. He is, "*From everlasting to everlasting.*" (1 Chronicles 29:10; Psalm 41:13; Psalm 90:2; Psalm 106:48) as each verse says.

THE ONES LEFT BEHIND

Fortunately for us, time is still running. It's taken all of the 13.8 billion years of the existence of the universe and 4 billion years of life on earth for us to get to where we are, as we are, who we are - and here we are, today. Voila! Which brings us to where science gets confused. How can there just be such a perfect *Goldilocks* world, not too much of this, not too little of that but just right - across as least 25 different critical physical constants.

Every one of those at least 25 critical physical constants must be set just so. If any were a tiny bit smaller or a tiny bit bigger the universe, let alone life itself, would not exist at all in what's known as a *"fine-tuned universe."*[55] See Also Glossary of terms.

Which brings us to the *"anthropic principle"*. There appears to be a kind of inevitability about the very structure of the universe whereby it, and ultimately human beings even exist. Yet, it is so very unlikely we could ever be so lucky. It has been said, *"It's almost as if the universe was waiting for us to happen."* The universe must be biased in some way but surely it was not created exclusively for our benefit - if there is no God, which to some degree confounds the anthropic principle. Then there would be no fundamental meaning to life. Admittedly, this is not an argument for God, although some scientists simply maintain it's meaningless to even search for meaning in life. From a purely scientific viewpoint, our brains, indeed our minds (two different things) are simply a collection of neurons sparking vast numbers of tiny sparks of electricity.[56]

Science instead asks, *"How could things ever get to be just so right for our very existence to exist unless there are a multitude, indeed an infinite number of potentially infinite other universes?"* The physical constants, indeed the very laws of physics, could be different. But if just one of them was different from what it is then most likely this would destabilise the rest to the extent we wouldn't ever have got here. However, it is logically impossible to have any evidence of an infinite number of infinite universes since we can never see beyond the one we're in, by definition - it being infinite.

To examine just one very commonly referenced, existential real-world constant. π is the ratio between the radius of all and any circle ever and its circumference. It is described in mathematics as an *"irrational number."* It has an infinite number of non-repeating numerals after the decimal place as in **3.14159265**... and so on and on - for ever. Not only does π relate to the square of the radius of the 2-dimensional space of circles but also to 3-dimensional spheres. It frequently and unexpectedly appears in many fundamental equations of the astrophysics of the very largest galactic proportions and the quantum physics of the very smallest. π, it could be said, is *"hard wired"* into everything -somehow.

If our universe is one which actually works by pure chance, and from its random selection of 'elements' we are all made, then arguably, a creator God is unnecessary. But such metaphysical universes can hardly qualify as science without any data. How these multiple other universes came into existence begs the same question, especially when they, like ours are considered to be infinite. An infinite number of infinite universes seems to me the reductio ad absurdum I mentioned earlier. It still seems absurd to believe a creator God does not exist given all there is in existence and the many peculiarites of the way it exists. However, although a distinct possibility, it seems not to be seriously considered in most scientific circles since the fundamentals of science rest on foundations of reproducible, digital data. Of course, if this is a Christian God of Love, which most of us would accept cannot be scientifically identified let alone measured, it becomes inexorably a matter for the individual, of personal choice. But it reveals the lengths to which one has to go to speculate on creation without a creator - which we could simply say is God. Might scientists just happily leave it at that as being beyond the limit of their remit?

Nevertheless, as the argument goes, a creator-God does not exist because of the moral conflict within and between religions. If this were a valid test of the truth of non-existence, then the very same argument could be applied to argue humankind does not exist because of moral conflict within and between communities. In both cases, does it not seem absurd as a means to establish the truth or untruth of existence - of anything?

More of this a little later, but all humankind's multitude of creation stories have one thing in common - a creator. It has been said, *"All religions lead to the same place by different routes."* The scope of this book does not allow me to venture into comparative religions. All of them have things in common e.g. a creation story of some kind. There are many more languages than there are religions, yet all of them have things in common e.g. a grammar of some kind. However I do favour Christianity's reconciliation of Love with Justice through the substitution Jesus made, his life in return for ours. Put this way *"God made him who had no sin to* **become** *sin (all that is wrong) for us, so that in him we might become the righteousness (all that is right) of God."* (2 Corinthians 5:21 TPT - my emphasis and italics)Well, that is what is says. Or, as C.S. Lewis put it in his book *Mere Christianity, "The son of God became a man that we may become sons of God."* At which point I agree, it might be a good idea to pause and make a cup of tea...

But believers in each religion would identify greater or lesser differences between theirs and the other ones. So significant are these differences; as mentioned earlier, that all including Christianity can be wrong, but they cannot all be right.

Back to Time and the Universe as We Know It.

Atomic clocks give an example of the order and consistency of the passage of time. Some use the constant frequency of oscillation of Caesium atoms for the tick-tock of a clock. This is so consistent and reliable that Caesium clocks can measure the rate of the passage of time with an accuracy of 1 second in 300 million years.

But even this extreme accuracy is not limited by any variation in the frequency of oscillation of Caesium atoms themselves. The measure of their accuracy is right at the limit of the best technology we currently have to check it. The calculated consistency of the oscillation frequency of ALL Caesium atoms, wherever they are, is actually less than 1 second in greater than the age of the universe! This is an expression also of the profound regularity of the passage of this created thing we call Time. Once more we are constrained to ask, "How can this be, all by itself?"[57]

And So to the Timeless

Returning to the subject of time itself, to explore further such things as the 'timeless particles' of which 'protons' and 'neutrons' are held together which quantum physicists refer to as 'gluons.' Not that we'd ever notice but these particles get around within each atom of which we and everything else are made, quite literally, at the speed of light. Just so, *"God is light."* (1 John 1:5) - of which, and so of Whom we are made. No, things are not at all as they seem.

The equally prosaically named 'strong force', holds all the 'protons' and therefore every atom in our bodies together, as well as everything else in existence. Just so, *"God is love."* *"He is before all things, and in him all things hold together."* (1 John 4:8, Colossians 1:17) - again, with Which and by Whom are we all held together? Yes, I know. But it's just the way it is.

Let There Be Light.

These 'elementary particles' (elementary not simple) of visible matter are held together by 'gluons' and the 'strong force.' They actually consist, at the most fundamental level, of just two so-called 'quantum energy fields'. As mentioned earlier, they are the 'quark field' and the 'electron field.' They literally permeate or *"fill the whole universe"* - as does Jesus who, *"ascended higher than all the heavens, in order to fill the whole universe."* (Ephesians 4 :10) The same Jesus *"for whom and by whom all things were made"* (Colossians 1:16, John 1:1-4) Well, that's just what it says.

Just so, *"God said, 'Let there be light' and there was light..."* the first day, the first *'yom'* or period in Hebrew of Genesis 1:2. This first *undefined period of time* became defined at 380,000 years after the, *"in the beginning God"* of Genesis

1:1. The afterglow of this primordial light or 'relic radiation' is still visible to radio telescopes today. It's known as the 'Cosmic Microwave Background' - a 1978 Nobel Prize-winning discovery by Arno Penzias and Robert Wilson.[58]

So, here it is, all this raw information. It doesn't really correspond much with our own perceived *reality,* such as it is. It is only *by faith* we believe what science tells us about the things I've just described of which we can have no actual knowledge. It's also, *"By faith we understand that the universe was formed at God's command so that, what is seen was not made out of what was visible."* (Hebrews 11:3) *"By the word of the LORD the heavens were made, their starry host by the breath of his mouth."* (Psalm 33:6) One of Christianity's Zen-like statements I can vouch for is, *"Don't try to understand in order to believe. First believe and then you will understand."* Not so different from many aspects of the science we've looked at here, not to mention of God too for, *"Now I know in part; then I shall know fully,* **even as I am fully known**" by Him. (1 Corinthians 13:12)

Therefore, *"We fix our eyes not on what is seen, but on what is unseen, since what is seen is temporary, but what is unseen is eternal."* (Corinthians 4:18) Once more this serves to remind us of Jesus for, *"In him, by him and for him all things were created: things in heaven and on earth, visible and invisible. He is before all things."* (Colossians 1:16.) But, as Jesus himself explains, *"I and the Father are one."* (John 10:30)

While people may say "God is in heaven," I would repeat, "No, heaven is in God." *"The Lord our God...... stoops down to look on the heavens and the earth?"* (Psalm 113:6) He is so much bigger than this.

If this God could indeed create and sustain the Universe and everything in it, protons, electrons, atoms, molecules, the planets, the stars, neutron stars, red giants, white dwarfs, super novae why then could he not "reveal his thoughts to mankind." (Amos 4:13) *"He hears ME when I call to him."* (Psalm 4:3) *"The Lord is MY shepherd."* (Psalm 23:1) This God who created the Universe and everything in it, seen and unseen "IS love." (1 John 4:8,16) and *"....whoever believes in him shall not perish but have eternal life."* (John 3:16)

Very soon we will explore even further what this *"love"* means in scientific terms and speculate about where and how this *"eternal life"* might be for us.

To continue, this *"heaven and earth" (Mark 13:31)* this universe, is not only expanding but it's doing so at a very precisely observed accelerating rate - balanced so to speak on something much finer than a knife-edge. The rate of acceleration is not quite so slow for the universe to have begun to collapse back again into a *'Big Crunch'* as it's known, under the inward pull of its own gravity. Neither is its rate of acceleration too fast for it to have disintegrated under the outward counter-gravitational push of the 'dark energy' mentioned earlier.

This *knife-edge* is known as the 'cosmological constant' featured in Einstein's 'field equation of general relativity.' I won't reproduce it here since it's even more obscure than the title, but it was proposed in 1917 to explain the then understanding of a static universe, one in perfect equilibrium with gravity. It was believed the universe was in a steady sate, neither expanding nor contracting so a 'cosmological constant'[59] was required to balance the pull of gravity in Einstein's equation.

Then, in 1929, Edwin Hubble discovered the universe is not static at all but actually expanding, requiring the 'cosmological constant' be set at zero. However, in 1998 the expansion of the universe was found to be accelerating. In some ways the 'Big Bang' is still *banging*. It's still happening albeit at a much slower rate than the original all but instantaneous 'inflation'

CHAPTER 13 - WHERE DID ALL THIS COME FROM ANYWAY?

immediately after the 'Big Bang.' Energy is still being created as 'dark energy' right here and now all across the universe. This knife-edge 'cosmological constant,' was reassigned a value as being the tiny margin between the 'dark energy' driven expansion and gravity driven collapse of the universe. It was found to be 1×10^{-52} which is so fine as to have 52 zeros after the decimal point. Here's just one measure of just how *finely tuned* the universe is for it to hold together - might one say, *"Just as it's meant to?"*

Here we have yet another example of how God does not hide things from us but hides things for us to find. If nothing else, it should serve to keep us in awe of who He is. We are at risk of becoming too comfortable in thinking our knowledge of *'what is'* in some ways exceeds the awesomeness of *'whose it is'*.

Might this relate to Adam's original mindset issue? As he was tempted to *"be like God"* by eating the fruit of the *"tree of knowledge"* (Genesis 3:5) so might we by focusing on the fruit of the tree (while others might be avoidng anything to do with it) instead of on its maker? Just wondering.

In any event, back to the chase. The concept of a 'fine-tuned universe' mentioned earlier, is where at least 25 'fundamental physical constants' are found to be set exactly to finely tuned Goldilocks or God factors. Neither too fast nor too slow, neither too large nor too small, they govern that many binary win-or-lose factors of existence with an accuracy beyond imagination.

If any one of them was wrongly set on God's kind of intergalactic dashboard by the smallest degree the universe may not exist and life certainly could not.[61]

Vast amounts of so-called 'dark energy' are continually being created, accelerating the expansion of the universe by just exactly over-compensating its accelerated collapse under gravity. Yet although crucial to our existence it's invisible to all the tools of modern science. There is a growing multitude of often contradictory scientific theories to explain what exactly 'dark matter' and 'dark energy' are. However, the very proliferation of so many theories bears witness to the importance and validity at least of their existence.

Even cosmologists really struggle with how all this is. Driven by a multitude of interlocking, interconnected super-finely tuned constants, how exactly did the universe come into and remain in existence as it is? That it did so at all is beyond doubt, except amongst a select few who speculate on whether consciousness has any meaningful existence at all. Was it simply by chance - because it did? Most scientists go to enormous lengths to avoid the conclusion, *"It must be God."* We lack a verifiable, repeatable, scientific process for this to be decide either way.

Although not without controversy, even within the scientific community, there's much hypothesising about multiple other universes or a 'multiverse.' Each one of an infinite number of infinite universes has different laws of physics and constants of its own.[60] But even these theories, especially not having any observable supporting data, do not solve the problems of by whom, of what or how they were made. Which takes us to what the scientific community refers to as "the First Mover" - or just back to God and how did He do it?

We might well ask ourselves, "Who am I to say a Creator God, although capable of making and holding all this together, knows and cares for me and reveals his thoughts to me?"

We might continue to ask ourselves, "Who am I to say a Creator God, capable of making and holding all this together, does NOT know and care for me and reveal his thoughts to me?" On what grounds?

At this point, amongst others, I have to say, *"Lord, I believe. How did you do it? How do you keep it all going?"*

Chapter 14

What are we all really made of?

Three quarks for Muster Mark!
James Joyce, Finnegan's Wake 1939

Returning briefly to Earth's atmosphere, made from a mix of approximately 20% Oxygen which we breath in and *burn* to provide us with energy. Much of the mass of plants comes from the air's 80% Nitrogen content. Fortunately for us they turn the Carbon Dioxide we breath out back to Oxygen for us to breath back in again. Neither this Oxygen, Nitrogen nor Carbon, nor any of the other elements heavier than Hydrogen and Helium were made at the 'Big Bang.' They were created hundreds millions of years later, and hundreds millions of light years from here. Yes, the mind can but boggle. Things are just not as they seem with our hot and cold running water, paved streets, internet, smart phones, supermarkets, coffee shops and so on...

All this, and all the matter in the universe ever created - all that of which we are made came somehow into being, as we know, in just 10^{-36} seconds 13.8 billion years ago. How do we know this? Light reaching us from the furthest reaches of the universe only just now began its journey 13.8 billion years ago in time. This means it began at a distance of 13.8 billion light years away from here, from the place it was put in the beginning. Yes, within the first 10^{-36} seconds of its creation.

The 'Big Bang' must have happened somewhere. Where did it actually happen from? Another great question. The universe went straight from nothing to everything in that 10^{-36} seconds. Nowhere became everywhere so fast there was no somewhere where it all came from. Just that, *"God said..."* and it was.

Having observed that everything in the universe is moving away from us it could seem we're right at the centre of it all, where it all began. Why here, when intuitively this seems so unlikely? Imagine space instead as the elastic surface of an expanding balloon or globe. Then everywhere is moving away from everywhere else at the same time. What's inside the balloon? Well, its surface, space and time itself, is all there is. There is no inside since space and time are constrained to its surface. The question, "What is this globe expanding into?" comes out the same. There is nothing around space and time, apart from God from where it originally came, to expand into.

Returning to the 'Big Bang', there were yet no atoms just then, nor even 'protons.' There was only a dense cloud or 'plasma' of the loose 'quarks' and 'electrons' from which 'protons' and 'neutrons' and therefore 'atoms' would be made. Just as described here in the Bible, everything was, *"formless and empty, darkness was over the surface of the deep"* (space). (Genesis 1:2)

Even having "cooled" to 10^{10} degrees it was still too hot for atoms to form until approximately 380,000 years later. The universe cooled sufficiently for this cloud of 'quarks'

CHAPTER 14 - WHAT ARE WE ALL REALLY MADE OF?

and 'electrons' to condense or coalesce rather like droplets of rain do from a raincloud into 'atoms' of the two lightest elements, Hydrogen, and a little Helium, nothing else. Then, like sunrise shining through the rain, the light of the 'Big Bang' finally shone through as, "God said, "Let there be light and there was light." (Genesis 1:3)

Over the next billion years or so, following what's known as the 'recombination epoch,' gravity began clumping together the dispersed Hydrogen and Helium gas into clouds. Further compressed by gravity, friction between the atoms heated them back up to millions of degrees. Nuclear fusion, by which stars "burn," began converting Hydrogen to more Helium. Nuclear Fusion converts matter back into energy, again according to Einstein's famous equation of $E=mc^2$. Rather like a vast, continuously exploding yet highly controlled Hydrogen-Bomb it powers the sun and stars to this day - and sustains all life on earth. *"And God said, "Let there be lights in the vault of the sky to separate the day from the night."* (Genesis 1:6)

Some of the lighter elements were created in the fusion heat of suns or stars. Some suns, tens to hundreds of times bigger than our own, are known as 'red-giants.' As their Hydrogen to Helium fusion fuel process eventually runs out, they go into catastrophic collapse under their own gravitational weight. A dramatic tipping point is reached when the outward force of fusion is overcome by the implosive force of gravity. Catastrophic collapse happens at 2/3 the speed of light creating a 'super-nova' as it contracts to a tiny fraction of its size. The remaining heavier elements were created in the ultra-high-pressure, super-heated temperatures generated during their collapse into supernovae. They in turn were explosively ejected at 99% the speed of light in every direction - including our own.

THE ONES LEFT BEHIND

First Atoms

CHAPTER 14 - WHAT ARE WE ALL REALLY MADE OF?

So it was. These 'elements' did not all become lost in the 'interstellar medium.' They gravitated together as dust and asteroids across distances of hundreds of millions of light years. They are the left-over ash or star dust - of which we are made. In our case it was attracted towards the sun, itself just an otherwise unremarkable star, one of billions in our own Milky Way galaxy, itself one of billions of others in the universe. Oxygen, Iron, Carbon, Hydrogen, Sodium and all the other crucial *"ingredients"* to life found their way here to our newly formed earth and even now into our very blood streams. Looking around us who would ever guess where they came from in our own experience? And yet, there they are for there are, *"More things in heaven and on earth than are dreamt of in your philosophy..."* (Hamlet Act 1 Scene 5).

This star dust or nuclear waste of which we are made accumulated on earth over billions of years through shower upon shower of asteroids. While there were still traces of Iron and the other heavier metals in surface rocks, most of them sank towards the earth's core. They include Silver, Gold, Uranium and so on. The lighter elements such as Carbon, Sodium, Oxygen and Nitrogen accumulated at middle and higher levels of what became known as the biosphere and the atmosphere.

The earth's core is kept molten by heat generated from the decay of radioactive elements it contains, like one over-sized nuclear reactor. These heavy metals had melted down to liquid form through the same frictional heat of gravitational compression. Quantities of each element gradually built up in quite gratuitous proportions to make what became the earth's surface sheen or *crust*. Others were released as gasses to form a flimsy atmosphere. The middleweight elements, like Carbon and other trace elements essential to life, accrued closer to or on the newly formed surface.

Fortunately for us, the upper atmosphere also contains an Ozone layer of three atoms of Oxygen or O_3 instead of the usual O_2 we breath. If all the Ozone in the upper atmosphere were compressed to the pressure of the air at sea level, it would be only 3 millimetres (1/8 inch) thick. But fortunately for us it's just enough to protect us from annihilation, from deadly gamma radiation from the sun. It also protects us almost completely from the constant bombardment of deadly cosmic rays originating at close to light speed from outer space. They would otherwise burn us instantly to a crisp if the Ozone layer was not there. Fortunately again for us, it is not stripped away by the solar wind of high energy radiation as happened on Mars. It clings on, gravity assisted - literally for dear life.[62]

This ubiquitous, somewhat mysterious, force of gravity holds our atmosphere and the Ozone layer umbrella in place. Not only is gravity pretty powerful in our own personal experience. Gravity is powerful enough for the moon to bodily lift the very oceans upwards by tens of feet towards itself. The moon's gravity prevails over the earth's pull on the oceans to this extent, despite the earth's much superior size and therefore gravity.

And yet, the so called 'strong force' holding 'protons' together, is 10^{41} times more powerful than that. Relatively weaker 'gravity' nevertheless acts right across inter-galactic space, while the 'strong force' is limited to sub-atomic space.

Meanwhile the Hydrogen and Oxygen atoms, two and one of each respectively, combined to make Water now found even in interstellar clouds of Water. Commonly found throughout the Universe, they arrived here most likely though bombardments of countless ice-bearing asteroids as Carbonaceous Chondrites containing up to 22% Water.[63]

Matter of all kinds, as solid, liquid or gas, contract as their temperature cools and the atoms lose energy. Fortunately

CHAPTER 14 - WHAT ARE WE ALL REALLY MADE OF?

for us however, unlike anything else, water somehow stops contracting as it cools below 4°C. At temperatures below 4°C Water instead begins to expand until it freezes solid at 0°C. Almost all substances contract as they cool. Water has the unique and peculiar property of actually expanding the colder it gets below 4°C, becoming lighter, less dense. This means as water cools below 4°C it actually floats to the surface of say any lake or pond or sea. As anyone knows who's been through a particularly cold snap, pipes and even heavy containers can burst as ice continues uncharacteristically to expand with literally tons of force the colder it gets. The slightly warmer water beneath finally freezes solid - from the top down. If this were not so, and water continued to behave like any other liquid to become more dense as it cooled and sank, then expanses of water would freeze from the bottom up.

Thinking this through, the early life which began in water would have been forced to the surface to freeze and die there instead of thriving beneath a heat insulating ice layer. Fortunately for us, life could never have developed as it has if water did not somehow behave in this way for some apparently unaccountable reason. All advanced life-forms like polar bears, seals, penguins living on the ice but fishing underneath it, could never have evolved. Neither could the fish in the first place. Did God actually design water to be like this, or just one more of those Goldilocks chances that facilitate Life and Evolution?

Like the universe itself Life needed a kind of '*Big Birth*' moment of creation for complex, highly accurately self-duplicating, self-correcting cells indispensable to life to just begin. This inexplicable mechanism of random mutation, natural selection and crucially perfect reproduction of the fittest took off about 4 billion years ago. It has been said that Evolution would have needed Evolution itself ever to get it started, to have come into being. Evolution couldn't have bootstrapped itself, but here we all are today.

If the random mutation process driving evolution over that period had been too slow, adaptations to changing conditions would have been too slow and life would likely have died out. If it had been too fast then successful mutations would have been overtaken by more mutations before they could become successfully established. Some would attribute this to the just right *Goldilocks* solution, others to - well, God.

So, billions of Carbon-based molecules, and other traces of the same supernovae-sourced Iron and so on which form the earth's core, circulate in our bloodstream to make and to deliver Oxygen to every cell in our bodies.

This is the host environment of who we really are - *"temples of the Holy Spirit."* (1 Corinthians 6:19) Tiny traces of Magnesium were also created somewhere else in the universe, maybe in our own galaxy. They are present along with other similarly produced elements which are key to the highly sophisticated machinery of photosynthesis. Plants and trees of all kinds convert the poisonous carbon dioxide we breath out in this way to produce not only Glucose for energy but also Oxygen - for us to breath back in again. Each 'molecule' of the active agent Chlorophyl consists of a highly complex relationship between 146 atoms.[64] They work together in an intricate process driven by light energy and heat from the Sun all of 93 million miles away as it is - that we may live.

It might seem arrogant at this point to assume we are the only intelligent beings in the universe, observing and comprehending it on whom God lavishes such love and attention - in real time. But there it is. How does he do it? Well, as unsatisfactory as it may sound, he's God. Things sounding too good to be true does not make them untrue. Just look at the *too good to be true* form human life itself takes. And yet, here we are.

We sometimes like to think life must have begun elsewhere in the universe, in a more noble way above and beyond this life

CHAPTER 14 - WHAT ARE WE ALL REALLY MADE OF?

here, described by Hobbes as, *"solitary, poor, nasty, brutish - and short."*[65] Somewhere, somehow life on other planets might approach more of our own high expectations of peace, truth, justice and wellbeing for all. One day, perhaps, we might ourselves evolve to be like this too. We might just make it. C S Lewis proposes we already have made it in his little book "Beyond Personality." There he points out, "Evolution itself as a method of producing change has been superseded. It isn't a change from brainy people to brainier people." It's a change that goes off in a completely new direction - from being creatures of God to *"sons of God,"* the *"new creation"* of 2 Corinthians 5:17. But I digress.

However, given the potential for life on other planets even to have begun and sustainably advanced ahead of us, none seems to have manifested itself here. We anxiously expect even now, to somehow discover extra-terrestrial life, traces of life on Mars or at least remnants of it. It goes almost without saying, we have not yet discovered any - but neither has it discovered us. To all intents and purposes, we're here *at home alone.*

Is there really *nobody else up there,* or is it almost as if by design? Can there really be a design without a Designer? Could it really all be by chance, notwithstanding evolution which could hardly have evolved itself any more than the universe did? I will return to these questions before finishing.

Made from Stardust - Nuclear Waste?

What a piece of work is man, How noble in reason, how infinite in faculty, In form and moving how express and admirable, In action how like an angel, In apprehension how like a god, The beauty of the world, The paragon of animals. And yet to me, **what is this quintessence of dust?**

Shakespeare - Hamlet Act 2 Scene II - 1599

To summarise. I don't know who you are, but I do know your body like mine consists of an exotic, finely balanced cocktail of 'atoms' assembled as elements, mostly Carbon-based molecules and compounds with a sprinkling of things like Sodium, and heavier metal-containing compounds. These fundamental building blocks of our lives are made from *elementary particles,* as in elemental, meaning primordial not simple. As we have seen, they are not even particles but swirls, eddies or vibrations in 'quantum energy fields' which are uniformly present across the cosmos. Giant suns, having imploded under gravity to become super novae, created them as by-products in their cores at temperatures of tens of millions of degrees. Billions of years ago, hundreds of millions of light years from here.

Now we marvel with Shakespeare, not fully realising quite, "What a piece of work is man?" We might well enquire with him about our being characterised as a "quintessence of dust." But how well put?

Different mixes of the same identical 'protons', 'neutrons' and 'electrons', make up every 'atom' in our bodies - a veritable "quintessence of dust." They're the very same ones which constitute the entire visible universe. Fortunately for us, as we have seen, all the 'atoms' of any 'element' are identical wherever we find them, wherever they came from - with not a single mistake ever having been discovered.

These 'atoms' are almost entirely totally empty space. There's quite literally nothing - at all - between the tiny 'nuclei' and the cloud of 'electrons' orbiting them. They're not just 99% empty space, but 99.9% followed by 19 more nines; empty space with nothing else at all in there. In other words we're almost 100% full of ... nothing.

It has been estimated, if you took the entire human race and somehow removed all the space from within their atoms, including the water, they would be compressed down to

CHAPTER 14 - WHAT ARE WE ALL REALLY MADE OF?

about the size of - a sugar cube. That's the density of neutron stars. Hundreds of times greater than the mass of our sun, collapsed down to barely 5 miles across, they rotate at close to 700 times per second. Their diameter shrinks as they collapse under their own gravity, so their speed of rotation dramatically increases and surface speeds approach one quarter the speed of light. How so? The same principle applies to a ballet dancer or skater pirouetting with outstretched arms. The action of simply bringing the arms close into the body causes the dancer or skater to spin ever faster.

Consider 'neutrinos' also originating from the core of the sun and from way beyond our galaxy. Tens of trillions of neutrinos pass through the human body every second - without hitting anything. Each has a mass of 1/500,000 of an electron and travels at close to the speed of light. In principle, neutrinos don't interact with matter but we know they are so small and the space inside the 'atoms' we're made of is so empty they don't hit anything.[66]

Consider again the electromagnetic force of repulsion between positively charged 'protons.' As for the magnets explained earlier it must be overcome by an even stronger force to hold them together, to stop the 'nucleus' from disintegrating. Scientists call this quite prosaically the 'strong force.' How it originated before the 'Big Bang' is not only unknown but, given the scientific principle that something cannot come from nothing, it's considered to be unknowable. The 'strong force' like the 'electromagnetic force' and 'gravity' - just are.[67]

Consider next that 'time' itself is foundational to all existence and therefore to science itself. Time began at the 'Big Bang.' Before then time did not even exist and neither did the laws of physics. i.e. No time, no science. If we can rely on science at all then in its very own terms there can be no explanation for the origin of the universe before time began. Not only is it unknown, it's proven to be scientifically unknowable for all practical purposes.

The 'Big Bang' was not an explosion in space and time. It was the explosion OF space and time themselves. Even the question "What came before that?" has little to no meaning since things cannot happen when there's no time for them to happen in. Hence being, "from everlasting to everlasting," God is the very definition of "before the beginning of time." All the timeless 'photons' of light, of 'quarks,' 'electrons,' 'gluons' are but themselves expressions of the very nature or essence of God - if you see what I mean? Tough isn't it? Makes my head hurt too. A conundrum to which we will return.

However, as Christians we do know this, "*God is Love.*" (John 3:16) "*In Him all things hold together.*" (Colossians 1:17)

This *strong force* is 10^{41} stronger than what we experience as 'gravity.'

Might this be a measure of the power of God's Love holding all things together? But if God were to ever stop loving, and "*holding all things together*" through the power of his love, through the 'strong force' at work in every atom in the universe - well, there wouldn't even be a Universe.

Who Then Are We?

> *God thinks, therefore I am.*
> *Not Descartes*

In the face of all this incomprehensible wonderment about the nature of matter, time and space it's quite natural to wonder who then are we? I could wonder "*who then was the person with whom I shared my every thought, my wife of nearly 50 years and - where is she now? Who indeed then am I who shared my every thought with her and now can no longer feel her touch - or even her presence?*"

CHAPTER 14 - WHAT ARE WE ALL REALLY MADE OF?

It is generally accepted, and as in 1 Thessalonians 5:23, we as human beings can be described as body, soul and spirit. "Soul" is quite a loaded word and can have a variety of spiritual meanings. So we might sometimes substitute for "soul" the word "mind" in this context. The body part is self-evident, the distinction between soul and spirit is less so.

For example, colour is a property of light. It is entirely subjective to the observer. Colour is not an objective experience. Until it is *seen* by something or someone; light itself has no colour. It's just a tiny, narrow portion of a vast spectrum of wavelengths of energy radiation, which happens to be visible to most living things on earth. The eye part of the body intercepts these wavelengths of light and thereby informs the brain of the observer quite objectively it has done so.

But it's in the brain that actual colour is subjectively perceived. The human body distinguishes between them and tells our soul or our mind it has seen colours. Lemons are not yellow, and neither do they taste like lemons until we interpret the light as having a lemony colour and other chemical data as having a lemony taste. Daffodils are not yellow until we see them as such. Our spirit then bestows all kinds of shades of meaning upon them as I might *"wander lonely as a cloud and see all at once a host of golden (in Wordsworth's case daffodils."*

Do animals have souls? Why not? They too sense the world around them as we do. The unique, distinguishing feature of human beings, made in the image of God unlike animals and all other creatures, is I would suggest the spirit, referred to here in 1 Corinthians 2:11-12. *"For who knows a person's thoughts except their own spirit within them? In the same way no one knows the thoughts of God except the Spirit of God.* I would say, this spirit we have uniquely in common with God since, *"God is spirit."* (John 4:24)

Beyond this subjective conclusion of *colour* there lies a *higher* level of subjectivity if we think of the spirit as giving meaning, not just to the lemon, or daffodil but to the whole context of our lives and of everyone else around us. It's our *"spirit within us"* which demands answers to those questions, *"Why me? Why now? What does all this mean? Who am I? Where am I going in my life? Why....?"*

Although the Bible refers to *"body, soul and spirit"* (Thessalonians 5:23) as being the whole person, one does not need to believe in Jesus to have a sense of our being as having these distinct parts. However, it's not until we do get to where the Bible says, *"God is spirit"* (John 4:24) that we might wonder, *"What do I have in common with God and him with me?"* To which I would echo what has been said, *"We are like him, although he is not like us."*

We are informed furthermore as believers, all but incredibly, that through this spirit part of our identity, *"God raised us up with Christ and seated us with him in the heavenly realms in Christ Jesus."* (Ephesians 2:6) Let us not be backward at coming forward on such a key point. Although we have done nothing to deserve it, from the moment we believe - it's a done deal; seated with God, in Jesus, in heavenly realms. Once more, *"In the Bible because it's true."* Yes, I still look it up sometimes just to check.

Charmaine and I are therefore together even now, in these *"heavenly realms."* Furthermore, we always have been, since the moment we came to believe Jesus is who he says he is, and we always will be. Here there is comfort indeed. You may take this as just the musings of a poor bereaved man seeking comfort in religion and relief from grief. But hopefully, if you followed my reasoning up to this point, this assumption should not change the conclusion from what is written in the Bible - it's in there because it's true. Rather, I think it's the state of bereavement which encourages us to venture to

CHAPTER 14 - WHAT ARE WE ALL REALLY MADE OF?

challenge scripture more deeply than before, and emerge with more of its spiritual nutrients.

To digress only slightly, I believe when we pray not from earth to heaven but from our being present in *"heavenly realms"* to earth, our prayers become even more *"powerful and effective."* (James 5:16)

But, and here we go, where anyway are these *"heavenly realms?"* What is it like being *up there* and what on earth are we doing *down here?* How, as some have put it, to live an *ascended lifestyle?* Or, to lead the way to a further discussion, is it really a matter of *up there* and *down here* at all? Or, hold on, given what's come before and is to follow, might they be one and the same place?

In the words of the song by Kim Walker-Smith,[67a] although not true because of the song, but the song being true because it expresses Biblical truth:

> *Heaven is here now*
> *He's all around us*
> *Heaven is Jesus*
> *He's the moments we meet*

These thoughts have occupied the hearts and minds of countless people, of all persuasions and none all around the world for millennia. Descartes addressed this question of what above all is fundamentally true? What can we really know for sure as ultimate reality? His answer has become a significant part of the bedrock of Western philosophy.

> *"I think, therefore I am."*

I don't believe it was ever intended to mean "I think therefore I made myself," as might be assumed. The Enlightenment nevertheless created modern Western culture which put "me" or "self" at the centre of our thinking, the individual as of paramount selfish importance. A similar mistake had been made by the early church until Galileo. They falsely placed the earth at the centre of the universe, thereby having a profound influence on everyday thought, leading to the widespread belief that faith is fiction.

Meanwhile the scientific community tends to refer back to the time of Galileo as a parting of the ways between the falsehood of faith and the truth of verifiable experimental data. To paraphrase, but not to exaggerate the scientific maxim, "If you can't measure it, it doesn't exist" leaves vast gaps in our knowledge of ourselves and the real world, gaps which cannot by their very nature be quantified or measured.

Remember, "Morning boys. How's the water today? What the hell is water?" We barely recognise that we "swim" in a medium of cultural norms we barely see let alone consciously respond to, and yet it dominates every aspect of our thinking and social behaviour.

So, follow through with me these subsequent steps in the light of what has come before.

I think therefore I am
I think, therefore God is
God is, therefore I think
God thinks, therefore I am

Are we then perhaps, along with everything else, figments of God's imagination? Not necessarily such a bad thing, but what on earth might it mean? Taking it hypothetically is hardly an option when we find ourselves in life-challenging circumstances?

What then are we really made of?

"And God created..."
Genesis 1:1 - God 10^{-49} seconds before time began

Returning yet further to the transient ephemeral nature of matter, what are we really, really made of? What are the internal visible and invisible universes, this *"water"* in which

CHAPTER 14 - WHAT ARE WE ALL REALLY MADE OF?

or in Whom, *"we live and move and have our being?"* (Acts 17:28) If you hadn't already, now really is fasten your seatbelt time. We tend to think intuitively of these subatomic particles as actual physical - particles. We think of them as tiny gritty bits of stuff, just too small to see or handle. And yet 'quantum mechanics' tells us otherwise.

Physicists have developed something rather unimaginatively called 'the standard model.' It describes every aspect of matter known to exist constituting trillions of galaxies and 10^{23} stars. So, let us just *"Lift up our eyes and look to the heavens: Who created all these? He who brings out the starry host one by one and calls forth each of them by name. Because of his great power and mighty strength, not one of them is missing."* (Isaiah 40:26) God is - at the very least - that big!

There also happens to be around 10^{23} electrons in just one milligram of matter. This is to say nothing at all about coincidence. But it's to say everything about the vastness of the creation we inhabit, while being just the twentieth part of all we know exists.

Can whoever created the stars and knows them each by name have any problem knowing each one of us tiny human beings by name, so much tinier than any star? *"Indeed, the very hairs of your head are all numbered."* (Luke 12:7)

No wonder only the Bible can explain itself, and it does - uniquely with help from the Holy Spirit:

Although,

> *"'My thoughts are not your thoughts, neither are your ways my ways,' declares the LORD. 'As the heavens are higher than the earth, so are my ways higher than your ways and my thoughts than your thoughts.'"*
> *(Isaiah 55:8-9)*

and yet,

> "He who forms the mountains, who creates the wind, **and who reveals his thoughts to mankind**, who turns dawn to darkness, and treads on the heights of the earth - the LORD God Almighty is his name."(Amos 4:13)
>
> "It is I who made the earth and created mankind on it. My own hands stretched out the heavens; I marshaled their starry hosts."
>
> (Isaiah 45:22)

Some claim with certainty it could not have been a "*whoever who*" .... created all this but a had to be a "*what-ever which*". Might they seem to wander towards speculation into "*not-ever*" created to resolve the "something from nothing" conundrum. But, can there really be such a thing as uncreated eternal existence apart from that which is God - by definition?

Back from the metaphysics to the science, all is explained in the 'standard model' with just three 'fields.' They are not fields as in growing grass of course, rather 'energy fields' namely the 'electron' 'quark' and 'Higgs' Fields. They combine and are acted upon by just four forces namely the 'electromagnetic' the 'strong' and the 'weak' forces and 'gravity.'

Before moving gently on it's worth saying 'gravity' after all is not strictly speaking a force, self-energised and continuously acting like the others. Gravity occurs rather as a reaction to a change of speed, or direction. As a car we're travelling in accelerates or decelerates we feel a force pushing us back in our seats or pushing us forwards against the seatbelt. As the car goes round a bend we feel a force to the side acting in the opposite direction to the bend even at constant speed. We feel the force resulting from the constantly changing direction. The sharper the bend we traverse, or the faster we accelerate, the greater the force we feel. As the change in speed or direction stops, so the force also stops.

CHAPTER 14 - WHAT ARE WE ALL REALLY MADE OF?

Gravity itself is such a force. It results similarly from the speed of our travel around a bend or around a curvature in what Einstein referred to as *'spacetime'*. Again, please bear with me when his name comes up. It was Einstein's genius to have discovered these things. Part of his genius of explanation is we don't need to be an Einstein to follow them.[68]

We are all familiar with living in the three-dimensional space of up, right and sideways. We do this all the time, every day, minute by minute - in the dimension of time. Einstein's *'spacetime'* is therefore made up of these three dimensions of space and one of time. Our everyday lives are therefore four dimensional. Any kind of space we inhabit, and we are perfectly familiar with this, can be described as up, right, and sideways at any given point in time - followed inexorably by another point in space and time, even if we don't move a muscle. Fortunately for us, although one may say sometimes unfortunately, time never stops.

To adequately describe when something happened we need to know both where it happened (in three dimensions) and when it happened. That's *'spacetime'*. Now, this will sound weird, but *'spacetime'* is curved or bent by the very presence of matter, by any kind of stuff be it the moon, the earth, the sun, the galaxies acting upon each other and on ourselves. We especially feel the effect of this force of the earth's gravity on our bodies as we get out of bed in the morning, or climb a flight of stairs. Consciously or unconsciously, and ever so fortunately it's with us throughout the day, every day.

In our case, this force of gravity we feel is created by the presence of the mass of the earth bending or curving *'spacetime'*. As we follow the bend around this earth-induced curvature in *'spacetime'*, we feel the same kind of force as we do when driving round such a bend in a car. In the former case we call it gravity. It's not acting sideways as in a car of course, but downwards towards the centre of the earth, which is itself doing the bending of '*spacetime.*' Gravitational pull is thereby created in our particular locality, which we perceive to be downwards.

We can measure this force with the most extraordinary accuracy as its effect permeates and interacts with every object in the universe - and never stops. Why and exactly how does it happen? We don't know, just as we don't really know why or even how electrons just keep on going round nuclei without ever stopping - for billions of years. Once having been set in motion electrons fortunately never ever stop no matter how cold it gets. Down even to what's referred to as 'absolute zero' of -273°C below which it's impossible to go - they keep going, forever.

Back out of the 'electron' rabbit-hole; the greater the mass, the more it distorts or curves '*spacetime*' the greater the force of gravity it exerts on other objects, as illustrated here.[69] Take for example the gravitational pull of the sun on the earth so it doesn't spin out of its orbit and off into outer space. The mass of the earth likewise acts on the moon keeping it in orbit around us.

Even light, which itself has no mass and is not therefore affected by the force of gravity, must follow the very curvature of '*spacetime*', created by all the massive objects in space. Nevertheless, it makes its way in time, over the billions of light years between them and us as observers at constant never-changing speed relative to everything else.

Gravity is the only phenomenon connecting the matter made of 'atoms' we can see and measure and 'dark matter', the other 70% we cannot. This, and 'dark energy', I will suggest later are quite literally the realms of heaven, the "*heavenly realms*" of the Kingdom of God.

Meanwhile the three 'energy fields,' although they too extend right across the universe in '*spacetime*', are different from gravity. They are actual fields of energy unlike gravity which

CHAPTER 14 - WHAT ARE WE ALL REALLY MADE OF?

is the result of a distortion of *'spacetime'* itself. The first two of these 'energy fields' produce matter in the form of electrons (from the 'electron field') 'protons' and 'neutrons' (from the 'quark field') which together constitute the 'atom'. Mass is attributed to matter by the 'Higgs Field'.[70] Without it, matter would have no mass. Without mass there would be nothing to interact with gravity as it bends *'spacetime'* thereby producing what we perceive as (our) weight.

I'll continue, to show how these three energy fields also relate to the Kingdom of God. Just as they permeate and actually are all the galaxies of the universe, the same energy fields permeate and are our bodies. In at least one sense these energy fields quite literally - are us. They are within you. Just so, *"the Kingdom of God is within you."* (Luke 17:21 NKJV)

Just as gravity and each of these three 'energy fields' extend throughout and fill the whole universe, Paul refers to Jesus when he says, *"He ascended higher than all the heavens, in order to fill the whole universe."* (Ephesians 4:10)

Returning briefly in this context to the invisible universe of 'dark matter' and 'dark energy.' They're referred to as being dark because they do not interact with light. Yet together they make up to 95% of the universe. We can neither see nor measure them other than by the gravitational impact they have on what we can see. We nevertheless, like Jacob at Bethel, do not even experience them here and now just as, *"Surely the Lord is in this place and yet we are not aware of it."* (Genesis 28:16)

Moving ahead a little further, the first of the three energy fields is described in 'field theory' as the 'electron field.' Progressing beyond the analogy for an 'electron' we're so familiar with (at least by name) as a fly buzzing around in a cathedral. They are generally visualised as some kind of physical particles, orbiting around the nucleus of the atom rather as planets orbit the sun.

Unfortunately, there are few if any valid comparisons to be made between the familiar world of Newtonian physics, and Quantum Physics. Gravity, as illuminated by Newton, has significance where masses are very great and distances very large. A body the size of the earth is needed to exert a gravitational force we can overcome by lifting a hand. A body the size of the sun is needed to exert a gravitational force sufficient to prevent the earth from spinning off into space, even from a distance of 93 million miles.

Newton's laws of gravity still govern calculations allowing us to control earth satellites essential to our daily lives, and to navigate spacecraft to the moon and outer planets. But Newton's laws simply don't work at atomic and sub-atomic levels. Gravity is negligible where masses are so minute and distances as unimaginably small at the micro-cosmic level, just as they are unimaginably large at the cosmic level.

So, contrary to popular imagination 'electrons' are not just tiny pieces of stuff orbiting nuclei pretty much like the moon orbits the earth. Instead of being discreet negatively charged particles or single points orbiting around a positively charged nucleus, 'electrons' manifest as very tiny waves or quanta of energy. The electron is a manifestation of the literally universal, all pervasive 'electron energy field.' The speed and position of each electron is indeterminate until measured. Even then only one metric can be known, either speed or direction, but not both. Electrons congregate as "probabilistic cloud layers" around the nucleus of atoms. Probabilistic insofar as their exact position at any moment in time within the atom is determined as a probability not as an actual location. In view of this 'uncertainty principle' the electrons within an energy level can be said to be everywhere all the time. Yes. In the quantum world the things of which we are made are weird!

CHAPTER 14 - WHAT ARE WE ALL REALLY MADE OF?

Oh, and they've all been doing this all the time since very very close to (though not actually quite since) the beginning of time, but just around 10^{-12} seconds afterwards. Since then, none of them have ever stopped moving. It's true to say there's nothing inside each 'atom' to stop electrons from moving since they were first created 13.8 billion years ago. It's also one of the unexplained features of atoms, their electrons never ever need extra input from an outside source of energy. And yet electrons never run down. Atoms never collapse or fall apart through loss of energy. These 'energy fields' of which we are made are constantly energised. They're in what's referred to scientifically as a *'quantum coherent state.'* In Biblical terms, *"God is before all things, and in him all things hold together."* (Colossians 1:17)

There is of course an electrical force of attraction between the positively charged protons and negatively charged electrons. Yet, they never crash into each other. But it's not the force of gravity holding them in equilibrium between their speed and their distance from the nucleus as it is for the planets orbiting the sun. As energy waves of strictly pre-determined unchangeable length, they are literally jammed tight into whole wave-length distances around the circumferences of their 'orbits'. These 'orbital circumferences' are therefore very precisely defined by the nature of their being.

These electron clouds form the *'energy shells'* which are the outer parts of the atom. As they come under pressure from outside, exponentially more and more energy is needed to compress them. So fixed and pre-determined are these whole wavelengths they'd literally need the energy of an atomic bomb to shorten and compress their *orbits* around the nucleus. Hence, the apparently hard surfaces we experience in the things around us like stones, bricks, diamonds and even the relatively firm boundaries of our own bodies. These materials may fracture and even decay into other substances, when we die, but it takes this huge amount of energy to alter their path around the atoms they inhabit.

So, where does all this leave us? With a very imaginative and powerful God, who spoke it all into existence in infinitely less time than the blinking of an eye, *"He breathed words and the worlds were birthed. 'Let there be,' and there it was - springing forth the moment he spoke. No sooner said than done."* (Psalm 33:9 TPT)

Yes, I know how you feel. I felt the same way when I discovered atoms would need an atomic bomb to crush the cloud of electrons shrouding them - even the atoms in the brain cells you're using to read this. When Rutherford first discovered this he was apparently, at least for a while, afraid to step out of bed for fear he might sink through the floor.

Where then did it all come from? We might be reminded, we have a God who *"calls into being things that were not."* (Romans 4:17) And so now they are.

But there is still yet more. Returning to the second of the three 'fields' named the 'quark field.' Yes, quantum physicists in the 1960s were a bit quirky especially in their nomenclature. Amend for clarity to read:

The word 'quark' comes from an obscure and somewhat whimsical quote taken in 1963 by the earliest developer of the concept, Murray Gell-Mann. He was awarded a Nobel Prize in Physics in 1969. It appears in James Joyce's, Finnegans Wake, *"Three quarks* (seagull cries) *for Muster Mark! Sure he hasn't got much of a bark"* whatever this means..... There are six types or *'flavours'* of quarks: up, down, charm, strange, top, and bottom, which you will be pleased to know we won't go into. Either way, the pronunciation of 'quark' is intended to rhyme with *"bark."*

Anyway, this nomenclature is quirky in a different way to earlier generations of scientists when as botanists multitudes

CHAPTER 14 - WHAT ARE WE ALL REALLY MADE OF?

of elaborate Latin names were preferred. But quantum physicists have rather fewer things to name as we'll see.

'Quarks' are tiny ripples, eddies or vibrations which manifest in the 'quark energy field' just as 'electrons' are tiny ripples, eddies or vibrations which manifest in the 'electron energy field.' Their names have little to no practical meaning although there again we will not go. They're simply names.

Nevertheless, all 'protons' are always made from one 'up quark' and two 'down quarks' while all 'neutrons' are always made from one 'down quark' and two 'up quarks'. 'Quarks' are the very building blocks of the nuclei of all the 92 naturally occurring elements. Our bodies, minds and even our memories, are made of 'quarks' as is everything else in the 5% part of our inverse we can see.

That's basically it. All matter everywhere in the visible universe, everything is made from just these three fundamental or elementary (although not simple) particles, namely 'electrons,' 'up quarks' and 'down quarks.' Each of them are manifestations of just those two fields namely the 'electron field' and the 'quark field.'[71] Yes, matter is as simple as this with regards to 'energy fields' ......in principle. I'm sure you've guessed there's more where we need not go for now without delving too much further into the so-called spin and so called colour of 'quarks.' Neither need we bother much with the fact that 'electrons' (not particles but ripples in an energy field) only have a probability of being at any one place at any one time. Meaning although confined to a particular 'shell' or energy level in a kind of cloud surrounding the 'nucleus', any single 'electron' inside an 'atom' is considered in quantum physics to be everywhere - all the time. Such is the nature of every 'atom' of which we, and everything else throughout at least the visible universe, are made. *Colour, spin, photons, anti-matter*, and *the myriad particle debris* produced momentarily when atoms are *smashed* etcetera,

etcetera can thankfully be studied elsewhere at leisure if this is what we want to devote any of it to.

To summarise, the three 'energy fields' have been compared with paint layered on the flexible canvas of *'spacetime'* curved and altered as it is by their combined impact upon it by mass. Might these universal so-called 'energy fields' be the very thoughts of God interacting as his brain-waves on a kind of cosmic ECG? No wonder science fiction revels in such things, but seriously.....!

Speaking scientifically, the seen 'energy fields' co-exist with unseen 'dark energy' and 'dark matter.' They exist in a framework of 'spacetime' moulded by gravity, which is enhanced by the gravitational effect of 'dark matter' itself. We can't see or measure either of them although they amount to as much as twenty times more than what we can see.

So far perhaps not all good. I recognise the challenge of suddenly being led beyond a zone of comfort I've taken quite some time myself to gradually explore. As yet we are so far removed from our everyday experience of who we are and what the world we see around us is. Not much, if anything, is as it seems.

But, at the end of the day so to speak, what holds or glues (hint) the 'quarks' together to form protons and neutrons? Another good question, to which you might recall the answer from before. Quantum physicists have named them as, yes 'gluons.' Well, why not? As massless particles they travel around inside their respective protons and neutrons (i.e. us) - at quite literally the speed of light - because they can and so they do. As we have seen and will explain further from Einstein's 'general theory of relativity' (no longer a theory but demonstrable fact) whatever travels at light speed can have neither mass nor be subject to time- just to recap.

Far more intriguing than their name, 'gluons' are therefore *massless timeless particles*. They always have been since -

CHAPTER 14 - WHAT ARE WE ALL REALLY MADE OF?

right, the 'Big Bang' - and always will be until, *"All the stars in the sky will be dissolved and the heavens rolled up like a scroll."* (Isaiah 34:4)

What will remain but, *"a new heaven and a new earth"* (Revelation 21:1) This I propose will be just part of the other 95% of 'dark matter' and 'dark energy' the Kingdom of God which is already all around us, co-existent, co-terminus with us even now, rendered visible to our spirit being when we die. Even now it is actually *"within us."* (Luke 17:21 KNJV) Well, that's just what it says.

But if we're all just made from 'energy fields' and 'massless particles' going around at the speed of light, where does what we experience as weight come from? *"Why are things so heavy?"* - as any 5-year-old might ask. *"What on earth is going on here?"*

So, we must press on to the question of 'mass' which gives 'matter' the weight we're all too familiar with when it interacts with 'gravity.' Here comes the third energy field. It's named after the person who theorised about it in 1964 and then helped to prove its existence in 2012. Like the 'electron field' and the 'quark field' the 'Higgs Field' referenced earlier permeates all the visible universe as well as all the rest since it somehow (nobody knows why) engages in the interaction between gravity and 'dark matter.'

The 'Higgs Field' imparts mass to 'atoms'. Without it the stuff of matter would have no mass through which gravity could impart 'weight' to our bodies and to objects we handle every day. No 'Higgs Field' no mass, no weight - at all. Without this so-called 'Higgs Field' well, there'd be nothing for the 'force of gravity' to act upon. Not only would everything float around, it would all fly apart - if it would ever have come together.

So, just to round things out, at least on 'energy fields' and 'mass,' this interaction between the 'Higgs Field' the 'electron field' and 'quark field.' This impartation of mass is mediated by an 'elementary particle' known as the 'Higgs Boson'. You may have heard about its discovery in 2012 (Nobel Prize 2013) having first been described by the same Peter Higgs in 1964. Here, with seat-belts re-tightened, we can embark on the four 'forces' interacting with these three 'fields.'

Back to gravity, the least well understood, or rather the hardly understood at all. Of course, we all experience gravity as keeping us seated on the chair we're most likely sitting on, or maybe the floor we're standing on. We experience gravity through weight, some of us having more of it than others. We know weight is the effect this 'force of gravity' has when it interacts with the 'mass' imparted to the 'electron field' and 'quark field' by the 'Higgs Field' through the 'Higgs Boson' where a 'boson' is basically a particle that carries a force.

Backing out from the inner "*particle cosmos*" to the "*outer space cosmos*" itself, every object in the universe is in motion relative to every other. I mentioned earlier our 100,000 km/hr annual safari around the sun at 82 times the speed of sound. The sun too is moving in relationship to other stars and galaxies. We are all travelling at some speed relative to everything else.

Speed is measured as the distance covered through 3-dimensional space (up, right, sideways) in a certain amount of time e.g. ourselves at 100,000 km in one hour. Einstein's uncontested view is this, both space and time are therefore inseparable in the continuum we mentioned which he refers to as *'spacetime'*. Gravity results from the distortion or curvature of *'spacetime'* wherever mass is present on the grand scale in the form of planets, stars and galaxies - but not atoms.

Of course, this is really very hard even for geniuses to comprehend, and it has been said there are very few people who ever have. Quite nicely stated but not at all well understood, "*Spacetime is curved by matter, and matter moves in directions determined by the curvature of*

CHAPTER 14 - WHAT ARE WE ALL REALLY MADE OF?

spacetime." It is what it is. *"God said, 'Let it be'..."* and it was.

In other words, *'spacetime'* becomes curved by the presence of 'mass,' which in turn produces the force of 'gravity.' As we move through curved *'spacetime'* the force of 'gravity' is again more like the force we experience when travelling in the fast car around a curve, a bend in the road.

To give a more graphic illustration. Imagine a heavy weight placed in the centre of a circular trampoline. It distorts the normally flat surface (*'spacetime'*) creating a smoothly conical depression in the centre of the trampoline. Imagine some small spherical objects thrown onto the trampoline. They will rotate or *orbit* around the central weight in ways mimicking the pull of 'gravity'. They will be *pulled* in towards the centre against an opposite outward force from the speed of rotation of the object around the central weight. This simulated force of 'gravity' itself varies according to both the distance from the weight and the degree to which the weight has induced a curvature to *'spacetime'*. In this case the curvature of the surface of the trampoline. The heavier the weight, the deeper and steeper the cone-like distortion or curvature of the trampoline, or the greater the simulated gravity. The further away a round object is from the centre while "orbiting" around it the faster it can travel. The more slowly it travels the closer it approaches and "orbits" the central weighty object while not coming close enough to crash into the centre.

One might well ask, *"But, end of the day, what does this really have to do with anything?"* All this must be accounted for when setting the atomic clocks referenced earlier on global positioning satellites. Earth's gravity out there, where the satellite is in orbit above the earth, is slightly weaker the further away it is from its surface. The curvature of 'spacetime' is slightly less since it's further away from the earth that's causing it. This means light has less far to travel around a straighter shorter '*curve*'. But since light always but always travels at the same speed relative to everything else, time itself passes slightly faster over that shorter distance since gravity is weaker. The onboard atomic clocks must therefore be set to run slightly slower than the clocks in our smart phones on the surface. Otherwise our observed position would be constantly adrift away from our actual position.

Furthermore, although light has no mass and can therefore travel at light-speed, its path is bent by gravity. How can this be? Another great question to revisit, just to be sure. Since gravity curves the '*spacetime*' in which light travels, then its path, too, becomes curved in following it. There's nowhere else for it to go. Where there is enough mass in one place '*spacetime*' becomes so distorted it curves right back on itself. The light being emitted from any so called *super-massive* object, becomes curved to such an extent it curves back on itself and cannot escape. There we have a 'black hole'[84] like the one at the centre of our own Milky Way galaxy. Sagittarius A* is no less than 4.3 million times the mass of the sun so it curves '*spacetime*' to such a degree light is bent back upon itself and cannot escape.[72]

Well, I didn't know much of this either until just a little while ago. But God's creation, how he did it all, and yet reveals it to us, is even more mind boggling than this. A world without the force of 'gravity' acting on 'matter' manifested by the 'electron field' and the 'quark field' through 'mass' imparted by the 'Higgs Field' through the 'Higgs Boson' is not only hard to imagine. I struggle to believe existence can exist at all like this all by itself, just like that, by some kind of accident. All we see around us, at the greatest and the least distances, is in such rigorous order and has such total consistency.

Which may sound pretty clever, for physicists to have discovered just how clever an Uncreated Creator God is to have created it and still be here sustaining it. In the words of

CHAPTER 14 - WHAT ARE WE ALL REALLY MADE OF?

the song, *"He never stops working. Even when we don't see him he never stops working."* Where and how? Since *"God is light"* (1 John 1:5) and massless 'gluons' move unhindered within us at the speed of light, God is there at the very heart of everything, always - working. Since *"God is love"* (John 3:16) he is there at the very heart of everything, holding it together with the 'strong force' - always working. If you would like to hear more on this please listen to David Tong's brilliant explanation.[73]

Seriously - How Did God Come Into It anyway?

'For in him we live and move and have our being.
Acts 17:28

At this point we may just want to ask, again *"Come on, where does God come into all this anyway?"* Such is His ubiquitous presence (timeless, massless gluons) the power of God's love (the strong force) and His Universal reach (the 'electron field,' the 'quark' and 'Higgs Fields') dear reader. The Kingdom of God truly is both everywhere at hand and here within us - or there is nothing!

Nevertheless, there might actually be a profound *nothing* or *nada*. Life without any meaning, does not make this "some**thing** or some**one**" of God true - especially as in the Bible, of course. And yet, *"the chief desire of humankind is not pleasure but meaning"* as is so powerfully depicted by Viktor Frankl in *"Mans' Search for Meaning."*[74] Who then are we for the universe to provide us with meaning? Who then are we in the face of this universe and why us? Might it come back to the enigmatic statement of the Colour Sergeant Major in the film Zulu? *"Cos we're 'ere lad."*

To reiterate Einstein's exclamation, *"The most incomprehensible thing about the Universe is that it is comprehensible."* Alternatively, returning to God as a possible anthropomorphic construct of our own. Have we made God in our own image? I would propose all the above shows it's far more likely God made us in His own image. Our image is to God what the self-portrait is to the artist. In other words, God is the artist, Jesus the model and we the portrait, each one of us His masterpiece. The art critic in this illustration might sometimes elevate himself to claim ownership of the studio, but no more. If asked of ourselves as the portrait, "What do we know about God?" the answer could well be, "Isn't He the guy with the paintbrush." Either way, although we are like Him, He is not like us.

Psalm 8

When I consider your heavens,
the work of your fingers,
the moon and the stars,
which you have set in place,
what is mankind that you are mindful of them,
human beings that you care for them?
You have made them a little lower than the angels
and crowned them with glory and honour.
You made them rulers over the works of your hands;
you put everything under their feet:
all flocks and herds,
and the animals of the wild,
the birds in the sky,
and the fish in the sea,
all that swim the paths of the seas.
LORD, our Lord,
how majestic is your name
in all the earth!

And yet, even so, we are all invited to become sons and daughters of this Creator, *"Majestic in name"* who, *"To those who believe in his name, he gives the right to become children of God - children born not of natural descent, nor of human decision or a husband's will, but born of God."* (John 1:12-13) Join me in contemplating how that might work and what it means for us.

Chapter 15

From Lowly Bodies to Glorious Bodies

Our citizenship is in heaven...the Lord Jesus Christ ....will transform our lowly bodies so that they will be like his glorious body.
Philippians 3:20-21

Jesus had just been crucified. The disciples were fugitives from the authorities, hidden behind locked doors. They must have known it. At any time, they could be arrested and themselves flogged and crucified, tortured to a long but certain death, just as Jesus had been. They awaited the same fate. Mary Magdalene had apparently seen and touched the risen Jesus by the grave, in the garden. The disciples had seen the empty tomb but as far as they knew the Master was dead and buried. Soldiers were expected to burst into their secret location at any moment to drag then away. (John 20:1-18)

Then all of a sudden, *"With the doors locked, Jesus came and stood amongst them."* John 20:19, 24.

Now, I don't think one of the disciples somehow slipped the door open unnoticed to let him in, or a kind of angelic Black Rod commanded entry. But there he was. Just like that. Jesus', *"Peace be with you"* was more than a casual greeting. *"Hey guys. It's me!"* would not have cut it. Not only had he been raised from the dead, he'd also basically just walked right through a door or even through the wall.

No wonder they were not at peace. On top of the events going on around them they must by now have been scared witless.

> *"They were startled and frightened, thinking they saw a ghost. He said to them, 'Why are you troubled, and why do doubts rise in your minds? Look at my hands and my feet. It is I myself! Touch me and see; a ghost does not have flesh and bones, as you see I have.'"*
> (Luke 24:36-39)

Jesus himself was standing there, in the flesh. Have you ever wondered how he did it? *"Well, he was Jesus,"* one might say. He could do anything. They'd seen him, *"heal the sick, raise the dead, cast out demons"* (John 14:12) and commanded them and us to do the same. But even by those standards this appearance was OTT. Even when he did it again a week later (John 20:26) Jesus', *"Peace be with you"* was still needed as more than just a greeting, especially for the previously missing Thomas. No spirit or ghost was He.

We don't need, nor can we necessarily have, scientific proof for such stories to be true and for us to believe. This story too is not true because it's in the Bible. Yes, it's in the Bible because it's true. While *"In the end it's not just the Bible in which we should believe but the One the Bible reveals,"* as Oswald Chambers puts it.[75] So, since it is true, and we do believe, science should not contradict but rather be aligned with scripture. Nevertheless, might there even now be scientific explanations for Jesus walking through walls, explanations which God has left for us to find? Such things are not hidden from us but hidden for us.

As Rowan Williams recently observed, "*It's about time to end the phony war between faith and science.*" But without some understanding of the kind of science we've been covering here it's hard to understand much of what's really going on in the Bible.

Even before his first visit to the disciples Jesus had to ask Mary not to "*hold on*" to him. (John 20:17) as, it seems she was doing. She somehow seems to have cohabited the same time, place and space as the risen Jesus.

What then was this dead and then resurrected "*flesh and bones*" body which walked through the wall, or otherwise just appeared? What then was this body, subject to the same 'gravity' holding them too to the floor, in which Jesus met the disciples in their own real time? In the past I'd assumed this was because Jesus in his "*glorious*" resurrected body was more real than the wall. However, it often got me thinking.

When Jesus first did this he'd asked, "*Hey, do you have anything here to eat?*" Well, it had been a rough few days to say the least. "*They gave him a piece of broiled fish, and he took it and ate it in their presence.*" (Luke 24:42-43) This really happened. The risen Jesus was actually hungry and needed something to eat. "*A week later his disciples were in the house again, and Thomas was with them this time. Though the doors were locked, Jesus came and stood among them. He still needed to say, 'Peace be with you!' Then to Thomas it was, 'Put your finger here; see my hands. Reach out your hand and put it into my side. Stop doubting and believe.'*" (John 20:26-27) Again they were invited to engage with this '*not a ghost.*'

Later still by the lake, *"When they landed, they saw a fire of burning coals there with fish on it, and some bread. Jesus said to them, 'Bring some of the fish you have just caught........ Come and have breakfast.' .... Jesus came, took the bread and gave it to them, and did the same with the fish"* (John 21:9-12) all in the *'real world'* in real time.

I think this is of much more than just passing interest. We are told, we too will one day have the same glorious bodies Jesus had then - and has even now. For He, *"by the power that enables Him to bring everything under His control, will transform our 'lowly bodies.' That they will be like His 'glorious body.'"* (Philippians 3:21) That's what it says. You can't make this up. There's no need to. *"For we know if the earthly tent* (body) *we live in is destroyed, we have a building from God, an eternal house* (body) *in heaven, not built by human hands."* (2 Corinthians 5:2)

Jesus's newly resurrected body is described in Philippians 3:21 as a *"glorious body."* It was no more the *"lowly body"* he'd had before, like the ones the disciples still had there and then. It was demonstrably no longer like the very bodies we have now which are so subject to decay and disintegration perhaps at a moment's notice.

The *"lowly bodies"* we have now will no longer be temporary *"earthly tents"* but the *"flesh and bones glorious bodies"* of the 35-year-old Jesus. Is this not something for us to aspire to, as Charmaine later did and now has? It will not be in this world, but since our *"citizenship is in heaven"* (Philippians 3:20) it will be in the *"far better place"* which most people intuitively know exists as *"a new heaven and a* (whole) *new earth."* (Revelation 21:1)

Supercalifragilisticexpialidocious! What else can I say and remain sitting.

CHAPTER 15 - FROM LOWLY BODIES TO GLORIOUS BODIES

Well, in these instances, Jesus's glorious body did not only inhabit God's timeless, *"from everlasting to everlasting."* (Psalm 41:13) Instead Jesus promises, *"I will come back* (in my glorious body in your real time) *and take you to be with me that you also may be where I am* (in your glorious bodies - in the same real time.)" (John 14:3) I say in the same real time because, as little as we do know about 'dark matter and 'dark energy' we do know they exist in the same real time and place as we do.

Meanwhile,

> *"This is no empty hope, for God himself is the one who has prepared us for this wonderful destiny. And to confirm this promise, he has given us the Holy Spirit, like an engagement ring, as a guarantee. That's why we're always full of courage." (2 Corinthians 5:1,5 TPT)*

So, where and how will this be? In accordance with scripture I believe this will be in our own newly glorious bodies like it is now for Jesus. This will also be in the same real time as Jesus was with the disciples - as we are now - in a kind of parallel world otherwise referred to as Heaven? Does it not follow?

THE ONES LEFT BEHIND

Chapter 16

When and How Will This Be?

> *In that day.......*
> *John 16:23*

So much for the *"lowly body"* we have now. It will become like Jesus's, a *"glorious body."* We will *"shuffle off this mortal coil"* and in that day be re-united with the spirit part of our *"spirit, soul and body"* selves. (1 Thessalonians 5:23)

As we know from before, having adopted the conventions that our *body* intercepts sensory signals of light and sound, their wavelengths and frequencies, while our *soul* perceives colour in the light. A lemon is perceived by our *spirit* having discerned the meaning of the lemon-ness of the light.

When we as believers *came to faith* back in the day, we took Jesus at his word. We believed he is who he says he is - both 100% man and 100% God at the same time. We believed who he said we are, sons and daughters of the Living God. Then in that day, in that moment, *"God raised us up with Jesus and seated us with him in the heavenly realms"* (Ephesians 2:6) as our spirit selves.

Well, this is what it says, past tense. It's happened already to our spirit selves, together even now, with Jesus, in heavenly realms. Without ever venturing anywhere near *"the coming ages"* of verse 7. What are we to do then about this outrageously bold so little discussed statement, *"seated in heavenly realms,"* laden as it is with mystery and hidden meaning for all believers, past present and future? We will return there as already hinted.

Praying then *"in the Spirit, from heaven to earth"* is another form of our more usual *"from earth to heaven"* prayers. For both, but especially the former, I'd go so far as to say it's impossible for nothing to happen in heaven when we pray in the spirit - since we are there already in the spirit. Thereafter, it's for us to contend for these answered prayers to become manifest here on earth.

Again, I might be forgiven for grief-filled wishful thinking, of being there already in heaven with my late wife. One may intone, *'The Bible doesn't say that. We don't go to heaven until we die. Everyone knows that!'* However, being *"seated with Jesus in heavenly realms"* is a simply stated Biblical fact. (Ephesians 2:6) He raised us up and seated us with him when we came to faith. Therefore, the spirit part of our unique "spirit, soul and body" existences must be there too, together in those heavenly realms, from the time we believed - even now. So they are! Our spiritual senses are just not generally aware of it. It's not part of our common everyday experience of course, although it should and can be.

For my own part, since Charmaine went on ahead to be with Jesus, I still feel left behind in the daily existence from which I write this. Nevertheless, ever since we *"came to faith"* Charmaine and I have been together in those *"heavenly realms"* for the past 25 years already, just as we are to this day. That hasn't changed. And so it is together in *"heavenly realms,"* for all believers, whoever were or ever will be. It might just be worth taking a moment of reflection here.

Hmm. But what exactly am I driving at here? One might well wonder, do our spirit selves exist simultaneously in two different places at the same time? Or is our spirit simultaneously present both here in earthly realms and in heavenly realms - at the same time? If so, might the earthly realms as we know them and the heavenly realms as portrayed in the Bible, the Kingdom of God in some way co-exist - together, as 'normal matter' and as 'dark matter' just as science maintains they do?

It's not our common everyday experience of course otherwise it would spoil the whole point of being here. And what, after all is that?

The challenge is this. We are quite familiar with the oft-quoted (Romans 8:28) *"....in all things God **works*** (my emphasis) *for the good of those who love him, who have been called according to his purpose"* as we all are since God, *"so loved (all) the world....."* But it begs the question, "What then is his purpose for us even being here?" I think it's fair to say we can rely on the Bible to explain itself, especially in this case by reading the very next albeit far less quoted verse 29, *"For those God foreknew he also predestined* (according to his plan - Phillips translation) *to be conformed to the image of his Son..."* In view of this common oversight in looking to verse 29 to explain verse 28 it's worth looking them up and comparing them with say 2 Corinthians 3:18.

Now, we know God himself is, *"from everlasting to everlasting."* (Psalm 90:2) He *"was and is and is to come, the Almighty."* (Revelation 1:8) He is outside time as a we know it. He created time itself beginning at the 'Big Bang' mentioned earlier, before which time simply did not exist. It all began 13.8 billion years ago from a *'singularity'* of zero size and vast energy - something which can be scientifically proven even though nobody can quite imagine it. And yet, when God as the resurrected Jesus was back in the upper room with the disciples his glorious body still coexisted in the same realm of time and place as their lowly bodies - in this our own *real time* realm as we know it even today.

We know too, this glorious body of his, will one day be ours, although for now, *"We long for our full status as God's sons and daughters - including our physical bodies being transformed."* (Romans 8:23 TPT) Nevertheless Jesus' transformed body was still somehow subject to the same gravity and the same passage of time just as were those of the disciples, as are ours, as are we - and as we will be. The resurrected Jesus did not just float around - he was right there, amongst them. Yes, he was able to appear and disappear, as he did with the disciples, walking along the road to Emmaus. (Luke 24:13-32) So in some ways Jesus's glorious body was like ours as we are, yet profoundly different in other ways, for scientific reasons we'll return to.

Instead of being in two places at once, might our enduring *spirit* selves actually be in the same place in the same realm of time, co-terminus in the here and now, simultaneously both on earth and in heaven?

Might it just be possible this earthly realm coincides with, coexists with the *"heavenly realms"* just as 'dark matter' and 'dark energy' do? Time for another breather, if you haven't yet thrown the book at the wall ....

"In that day" (John 16:23) the mystery will be resolved since Jesus promised, *"I will come back and take you to be with me that you also may be where I am."* (John 14:3.) He comes back for those who then go on ahead of us with him, as he did for Charmaine. I imagine a conversation between them on one of their frequent walks together. *"You know what, Charmaine?"* Jesus says. *"I've just noticed we're closer here*

to my place than yours. Why don't you just come on home with me?"

Later he will come back for me too. Now, as one left behind *"I know in part; then I shall know fully, even as I am fully known."* (1 Corinthians 13:12) I will be fully united with Jesus - with Charmaine, and others too, there where they are now.

Yes, I will explain a bit further what science has to say about this. We're actually getting there.

Where is there?

'Dark energy' and 'dark matter' constitute 95% of the total mass-energy content of the Universe.

> *"The Son .... holds the universe together and expands it by the mighty power of his spoken word.*
> *(Hebrews 1:3 TPT)*

Where exactly is this better place then, Wordsworth's *"very heaven"* we all intuitively long for? Isn't this a question everyone is bound to ponder at one time or another? Might we too exclaim as we go on ahead with Jesus, as in Revelation 21:3-4,

> *"'Look! God's dwelling place is among the people, and he dwells with them. They are His people, and He himself is with them and is their God. He has wiped every tear from their eyes. There is no more death or mourning or crying or pain, for the old order of things has passed away.... making everything new.'"*

What might science be telling us about this *"glorious body"* like the one Jesus has in these *"heavenly realms"* the same one we in turn can look forward to?

If you're still here, you might be thinking, *"Better not ask. Only believe."* Or, maybe it's time just to take another quick breather, process these questions and look back to what we covered in Chapter 12 - *The Kingdom of God is Within Us.*

A discussion has been maturing since 1884 about so-called 'dark matter' and 'dark energy' being 95% of all the matter and all the energy in the universe, dark because invisible. The stuff of Nobel Prizes not science fiction. It's known they co-exist nevertheless here and now with the light or 'ordinary matter' and 'ordinary energy' - if you can begin to see further where I keep going with this? All the trillions of galaxies and 10^{23} stars are but 5% of everything in existence in the entire universe. There's twenty times more we cannot investigate by any means known to science.

'Gravity exerts a pull on 'dark matter' just as it does on our 'ordinary matter' because it curves or distorts 'spacetime' in just the same way as 'ordinary matter' does. The gravitational pull produced in this way by what cannot be seen acts powerfully on what can be seen. This is the only way we know it's there since it's totally invisible to every other sense and to every kind of instrument we possess.

Back to how do we know this. As mentioned earlier, the arms of a galaxy rotate right across at the same angular speed as the centre. Every part keeps up with every other part. The arms do not fly off into space. This means, just like with the horses on a carousel, the centre turns very slowly while the rest turns ever more quickly the further out you go from the centre. The arms at the outer reaches of a galaxy travel fastest of all just as do the horses on a carousel. A carousel stays held together by wooden or steel structures. Galaxies stay held together by the gravitational pull of much more matter, by 'dark matter' which can not be seen or observed.

More 'dark matter' is needed to hold things together at the faster moving edges of a galaxy than at the slower moving centre. This creates a kind of halo or doughnut effect. The further from the centre the amount of gravity-creating 'dark matter' gradually increases.

CHAPTER 16 - WHEN AND HOW WILL THIS BE?

But if this 'dark matter' and the normal matter of which we are made, co-exists together with us in the here and now, why don't we feel it? Great question! Our earth is so very tiny, a few light seconds across by comparison with the size of the Milky Way of 100,000 light years. It follows, since 'dark matter' is spread across the entire galactic disc of say our own Milky Way, its distribution or density is low by comparison with the density and the gravitational pull of the earth.

Lord Kelvin suspected this back in 1844. His *unseen matter* was later termed simply *"dark matter."* The phenomenon of constant angular velocity therefore faster speed towards the edges of galaxies and slower towards the centre, was first observed in 1933 around the Coma Galaxy Cluster about 320 million light years from here.[76] Back then we could not observe our own Milky Way galaxy from the inside. Coma is a cluster of over one thousand or more galaxies somewhat similar to our own, each of which is at least 150,000 light years across.

The galaxies in the Coma Cluster were discovered to rotate at speeds indicating many times more matter is present there than can be seen by any means at all. Although invisible, the gravitational effect of 'dark matter' holds them together.

In later years, this phenomenon has been observed in our own galaxy as in vast numbers of other galaxies. Without the presence of such 'dark matter' throughout the entire universe primordial clouds of Hydrogen and Helium gas might never have fused into stars which later coalesced into galaxies as they did.

We and our solar system, situated towards the outer edge of our own Milky Way galaxy, whiz round its centre at 800,000 km/hr, although even then it makes just one rotation every 200 million years. It's a long way even at this speed! If it were not for this uneven distribution of 'dark matter' (more prevalent towards the edges) co-existing with our light or 'normal matter' our solar system, along with us and our little droplet-of-molten-magma-planet-earth, would spin off into outer space - but we don't.

In sum, 'dark matter' co-exists with us - in the same space and time we are in. It's with us right here and now, as if earth were permeated by or even within - well, *heavenly realms*, invisible but everywhere and ever present with us.

Returning to the similarly Sci-fi sounding 'dark energy.' It too is invisible and undetectable, but we know it's there because the universe is not only expanding its expansion is

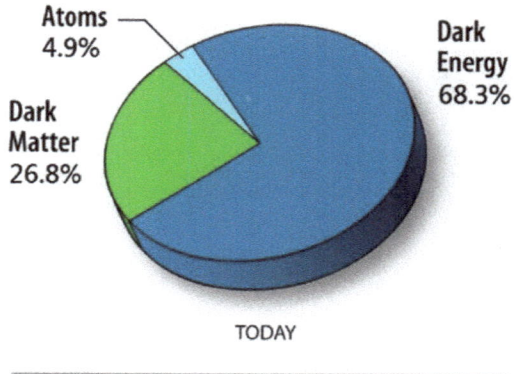

TODAY

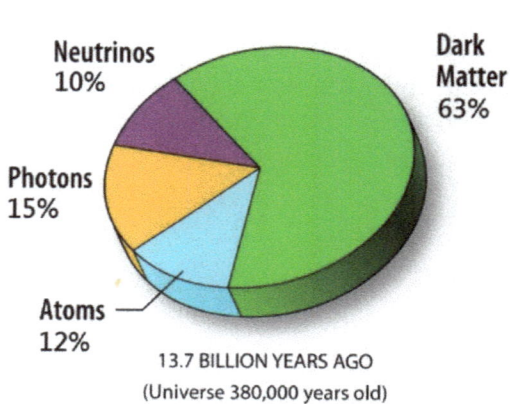

13.7 BILLION YEARS AGO
(Universe 380,000 years old)

Dark Matter Dark Energy

CHAPTER 16 - WHEN AND HOW WILL THIS BE?

accelerating. The continuous creation of 'dark energy' drives the universe to expand at an ever faster rate. Well might we ask, "What is the 'something' of the universe expanding into?" The answer in a word, more of the same so-called 'nothing' of God who was there before the 'Big Bang'.

This invisible energy is constantly being created and infused into the visible Universe. It's nothing at all like the vast amount of energy converted directly to matter at the instant of the 'Big Bang'. Nevertheless, it's happening quite uniformly, everywhere.

The expansive *push* of 'dark energy' just slightly overcompensates the contractive *pull* of 'gravity.' This small net acceleration is represented by the 'Cosmological Constant' as lambda (λ) in Einstein's 'Field Equations' supporting his 'General Theory of Relativity.' Calculated at 1.1056×10^{-52} (this is a very tiny number) but it's all it takes to make all the difference between a universe continually expanding from the 'Big Bang' instead of contracting down to a so-called '*Big Crunch.*'

To summarise, the extra gravity emanating from 'dark matter' is what holds ours and all the other galaxies together. 'Dark energy' very slightly over balances gravity, just accelerating instead of just decelerating the expansion of the universe. Or, as the Bible explains God both, "*.... holds the universe together and expands it by the mighty power of his spoken word.*" (Hebrews 1:3 TPT) Well, I didn't make that up. It's just what it says.

Might it be a logical option amongst other, in answer to the grand question, "*What or who was there before the 'Big Bang'*" to say "*God is by definition the One who created everything from nothing. God created all that is, just exactly as it is.*" By that simple definition God IS everything. Jesus, and the God of the Bible become almost (although not quite in my own mind) a matter of choice.

Just to recap a little further. In Chapter 12 - *The Kingdom of God is Within Us*, I explained how all the atoms in our body are 99.9[19] % empty, as in nothing else physically exists in there - at all. Could this cathedral-like space within us be Pascal's, *"God-shaped vacuum"* waiting as *"temples to the Holy Spirit"* to be filled by Him when we believe? (1 Corinthians 6:19)

These sub-atomic particles making up our bodies, our minds and our very memories are not just tiny gritty particles, smaller versions of those which get stuck in your shoe. Atoms are actually tiny disturbances or ripples in the 'electron field' and the 'quark field' which I mentioned earlier, permeating the entire universe. These ripples of massless energy only apprehend or acquire mass through their interaction with the so-called 'Higgs field' by means of the now famous 'Higgs boson.' These fields permeate the Universe, just as Jesus is *"the very one who ascended higher than all the heavens, in order to fill the whole Universe."* (Ephesians 4:10) But *"Why bring God into everything like this?"* I can hear the cries. How can we keep God out of it when he is so manifestly present in the natural world and put *Nature* as a place-holder kind of explanation?

Of course, we can no more experience 'dark matter' and 'dark energy' as human beings than we can experience all the signals and vast amounts of data emitted from our routers to our laptops, or the multiple satellites in orbit above the atmosphere transmitted to our smart phones. No, but as spiritual beings God has perfectly designed and equipped us, like spiritual *'internet routers'* to experience His presence and His love.

In this *digitised space age* characterisations of 'dark matter' and 'dark energy' can still seem quite strange to say the least. Or maybe now, if you've read this far, it seems even more so?

CHAPTER 16 - WHEN AND HOW WILL THIS BE?

And yet, we are invited to become no less than sons and daughters of the living God, of the very One who created all 100% of the Universe, not just the 5% we really know about. Referring to Jesus,

> *"He came to that which was his own, but his own did not receive him. Yet to all who did receive him, to those who believed in his name, he gave **the right to become children of God**— children born not of natural descent, nor of human decision or a husband's will, but **born of God.**" (John 1 11-13)*

"*Not true!*" some may say. "*Where's the evidence?*" But it's hard to deny this is an audacious proposition to be carefully examined, as profound and amazing as it is in the context we have been discussing.

As believers, we get to be adopted into God's family, as brothers but also even as friends of Jesus. "*I no longer call you servants, because a servant does not know his master's business. Instead, I have called you **friends.**" (John 15:15) Some may say, "*That can't be true given all you've said about the universe and creation*" but none can deny this is outrageously audacious and deserving again of careful examination. Who's to say it's not true, given a lack of evidence to the contrary? And, what if it actually turns out to be true, looking ahead to the day?

But this is not all. Jesus says not only to the disciples but to each one who believes, "*I am in my Father and you are in me and I am in you.*" (John 14:20) This sounds so totally, wonderfully amazing, but what on earth are we to make of it!? What does it really mean!? For when Paul says, "*Christ lives in me*" (Galatians 2:20) it's true for every believer, not just for himself. This Creator God actually comes to inhabit our '*lowly bodies*' even as we live and breathe in this world. Chosen to be so indeed "*before the creation of the world.*" (Ephesians 1:4) He co-exists with us, stands amongst us even as the resurrected Jesus is co-terminus with us, and within us - so He is!

Therefore, Paul speaks for all who believe when he says, *"Christ lives in me"* (Galatians 2:20) and not just in Paul since Jesus *"died for all."* (1 Corinthians 5:15) This is one of the very cornerstones of Christian faith, how is it possible? We don't need to know how this works in order to believe it, but having read this far, I think we can legitimately ask how.

Jesus' *"glorious body"* (yes, the same one we will have) his 'dark matter' presence was subject to both gravity and time. He made it known to everyone around for well over a month. So there I said it. Here's where you have permission to join me in falling off my chair - again.

In this science-corroborated faith context the idea of earth cohabiting heaven, as do 'ordinary matter' and 'dark matter,' it should not seem so absurd. Will not our 'ordinary matter' bodies, the ones we know so well and use to read this, one day be left behind to disintegrate? Then will not our spirit, our real enduring selves, the ones we know that we know that we know with, continue to inhabit those *"heavenly realms"* of the Kingdom of God? As we do even now?

We as the *"spirit"* part of *"spirit, soul and body"* will transition from this *"lowly"* 'ordinary matter' body to our *"glorious"* 'dark matter' body as we pass from mere life to eternal life. Will the other 95% of the true vastness of the Universe - the *"new heaven and the new earth"* - then become visible to us as our new home, where lies our true, heavenly *"citizenship."* (Philippians 3:20, 21)

Does this not coincide quite well with Job's insistent belief, *"I know that .... after my skin has been destroyed* ("lowly" 'ordinary matter' body) *yet in my flesh* ("glorious" 'dark matter' body) *I will see God; I myself will see him with my*

CHAPTER 16 - WHEN AND HOW WILL THIS BE?

own eyes - I, and not another. How my heart yearns within me?!" (Job 19:25-27 NIV) How indeed!

In this same context we can join Peter in repeating in the highest,

> *"Praise be to the God and Father of our Lord Jesus Christ! In his great mercy he has given us new birth into a living hope through the resurrection of Jesus Christ from the dead, and into an inheritance that can never perish, spoil or fade. This inheritance is kept (even here and now) in heaven for you."*
>
> *(1 Peter 1:3-4)*

Yes, that's what it says.

On Earth as it is in Heaven

As we know, Jesus told us to pray, *"Let your will be done on earth as it is in heaven."* (Matthew 6:10) It can seem to say, *"Let things work out one day and get done your way on earth as it gets done in heaven."* This will happen anyway. God will ultimately have his way this way in the end, whether we pray for it or not. More to the point, I believe it means, *"Let your will be done* (here and now) *on earth as it is* (as things are even now) *in heaven."*

However, in the here and now, there are endless medically verified examples of Jesus healing sicknesses of all kinds in today's world. He is *"Jehovah Rafah."* *"The God who heals"* is one of his names.

In common everyday language we use the word 'heavenly' especially in advertising as shorthand for all that's good. I think most of us would agree, at least in principle having assumed Heaven exists, it's a place where there's no sickness. In these healing events, Heaven invades earth and displaces the symptoms and root cause of the sickness.

Might this be where timeless particles of 'dark matter' are transformed into healthy 'ordinary matter' particles instead of say cancer cells? Might these instances of healing become in this way, "*on earth as it is in heaven*" i.e. no cancer here either?

But healing prayer doesn't always work out quite this way, of course. This is not the place for a dissertation on healing. But, for Charmine eight years of medical opinion were confounded by the off-the-scale results of their spirit-assisted medical interventions. Then, Jesus chose to take her home with him to be totally healed. Meanwhile, as before, it had been impossible for nothing to happen in heaven when we prayed from there, seated with Jesus in those "*heavenly realms*" (Ephesians 2:6) co-existing here and now with our place on earth.

Jesus's name was on the promise of healing, on the cheque so to speak. "*I will do whatever you ask **in my name**, so that the Father may be glorified in the Son*" (John 14:13) as Charmaine found it to be.

Which puts me in mind of a story from the old days of the British Raj. In India, a British army officer had only to sign his name on a scrap of paper addressed to his bank, "*Please pay the bearer £x.*" Such battered scraps of paper would apparently turn up at London banks years later, having been confidently transacted from hand to hand at face value, for the bearer to be paid the equivalent £x- in full, in his name.

Likewise, prayers prayed in the name of Jesus are *cashed* here on earth when heaven does invade earth from time to time, from place to place we will, "*See the goodness of the Lord in the land of the living.*" (Psalm 27:13)

Or, as later on for Charmaine they're redeemed "*there*" in heaven, having earned interest, so to speak. Meanwhile, expressed in the words of the song:

CHAPTER 16 - WHEN AND HOW WILL THIS BE?

Take courage my heart
Stay steadfast my soul
He's in the waiting
Hold onto your hope
As your triumph infolds
He's in the waiting[77]

As we begin "to know fully as we are fully known".

THE ONES LEFT BEHIND

Chapter 17

God is Light

*"We know in part; then we shall know fully,
as we are fully known even now."*
1 Corinthians 13:12

Some things we read in the Bible are so familiar we tend to bump over them each time and read on, so here's an opportunity to recap on something we touched on in several places before. In just three words we're told something so significant about who God is, yet we tend not to notice when it says, *"God is light."* (1 John 1:5) This is due at least in part because we must wonder at least subconsciously, *"What does it really mean anyway?"* Since both science and theology are true, science should and does inform theology and actually relate light to God.

As we know, time began at the 'Big Bang' from a 'singularity,' a point of *zero size*. Since energy takes up no space it contained all the energy ever needed to be converted in an instant into the entire mass of the universe. Before that instant in which time began there simply was no time, no space - just God. What or who else, again by definition? Time and space - *'spacetime'* did not exist. But, *"What came before time and space began?"* we're still bound to ask?

Back to the *user manual,* the Bible. *"In the beginning God created..."* (Genesis 1:1) He was there already, before the beginning. Many will debate the meaning, but this is what it says, *"From everlasting to everlasting you are God."* (Psalm 90:2) Again, in the Bible because it's true. God is outside time. God created time just as he created the universe and life itself from which evolution could evolve.

God is therefore timeless, beyond time. He, *"knows the end from the beginning."* (Isaiah 6:10) But even *"before the beginning of time"* (2 Timothy 1:9) God was thinking of us and making preparations for us to join him through *"the hope of eternal life, which God, who does not lie, promised before the beginning of time."* (Titus 1:2)

This is how it's possible to say of God *"You have gone into my future to prepare the way, and in kindness you follow behind me to spare me from the harm of my past."* (Psalm 139:5 TPT) What are we to say to this but as in the next verse, *"This is too wonderful, deep and incomprehensible! Your understanding of me brings me wonder and strength."* (Psalm 139:6 TPT)

We can no more fully comprehend a time before time than we can fully comprehend God, and yet there are things about both we can comprehend.

Indeed! How so? Einstein's original 'special theory of relativity' is no longer a theory but fact verified since 1905 in very many ways. I will enlarge a little more later on the examples I've given of how this theory is applied in our everyday lives in our smart phones, GPS satellites, internet routers etc.

CHAPTER 17 - GOD IS LIGHT

Put quite simply but profoundly 'simple relativity' tells us light always but always travels at the same speed relative to any object. However fast something travels, the light coming from any object will always arrive at another object at the same speed, the speed of light i.e. 186,000 miles per second, as we know. While being true regardless of how fast one object travels relative to another, it may still sound counterintuitive. It completely is so.

Imagine with me a car travelling at 60mph behind another one going at 40 mph. The faster car is approaching the slower car at a relative speed one to another of 20mph (as in 60mph - 40mph = 20mph). The two cars are travelling at 20mph relative to each other. Of course, the faster car will eventually draw level with and pass the slower one, at a perceived relative speed of 20mph.

If the faster car, the one behind, accelerates from 60mph to 70mph while the first car is still only going at 40mph, it will now approach and pass the slower car even faster at 30mph (as in 70mph - 40mph = 30mph) instead of just 20mph. The two cars are now travelling at a speed of 30mph relative to each other.

This simple principle of relative speeds applies to everything in existence - except light. Light emitted by one object travelling behind another will still be seen by the one in front as approaching it at the speed of light - regardless of how fast or slow one object is approaching another. This will be true looking either ahead or behind! Light from the car ahead reaches us at the speed of light too. It doesn't matter what speed we or they are travelling at. People travelling on both objects would observe light arriving each from the other at the speed of light at the same 186,000 miles per second, irrespective of their relative speeds.

So much for the relative speeds of cars. We might well conclude correctly, however fast or slow they may be travelling relative to each other (be they aeroplanes, satellites, planets, or stars) this way of calculating relative speed is true for all objects, everywhere - except for light.

At 186,000 miles per second or nearly 67 million miles per hour light could, in less than the blink of an eye, travel eight times back and forth from say London to New York. This constant speed of travel, regardless of the speed of the object emitting light relative to any other, does have some profound implications for the meaning of time, of who God is and who we are.

All the atoms in our bodies and the rest of the entire visible universe, are made of 'protons' and 'neutrons.' As we know, they themselves are made of 'quarks' held together by 'gluons' moving between them at the speed of light. As mentioned earlier, whatever travels at light speed is effectively timeless - just so *"God is light. From everlasting to everlasting."* (1 John 1:5 and Psalm 90:2) as are the 'gluons' holding us together!

Contrary to the manner in which they're generally characterised these 'quarks' are not tiny particles at all. They're not just little gritty bits of stuff, of the kind you might get in your shoe. They're merely those tiny ripples or eddies in the 'quark field' which fills the whole universe. Speaking again of Jesus how he, *"ascended higher than all the heavens, in order to fill the whole universe."* Ephesians 4:10) Here again, we find both scientific and biblical facts coincide.

From the external 'cosmos' we can see into the internal 'particle cosmos' a manifestation of a timeless God - which is what we are made of, both physically and spiritually. Yes, we are, which takes *"made in his image"* to a whole other level.

Now this part will take another bit of a stretch of the imagination, and thank you for staying with me this far. Since the speed of light is always the same, regardless of the speed of the light-emitting object, time itself in the vicinity of such an object must pass ever more slowly as its speed increases.

CHAPTER 17 - GOD IS LIGHT

Following this to a logical conclusion; for whatever object actually traveling at the speed of light, time in the vicinity of the object, stops altogether. It becomes timeless, outside of time, again like God. "*But why? How? Prove it!*" I can almost hear you say.

To delve just a little further into the science, take the example of two objects, say a stationary observer this time, to make it easier, and a moving object. Their relative speed in this case is simply the speed of the moving object. Each object has its own clock in this example. Time is measured for each one by the ticking by of time, whether tracked by a clockwork, or by a super accurate atomic clock. The relationship between ticks of each clock and the speed of the moving object relative to a stationary observer is represented by the following equation relating to what's known as Time Dilation. It's actually derived from some basic geometry and algebra. Still, not for the faint hearted but you can fast forward to the 12 minute mark of the Time Dilation video to see they do get there, and more if you're interested.[78]

t = *time between clock ticks for a stationary observer*
t^o = *time between clock ticks on the moving object*
v = *speed of the moving object*
c = *speed of light*
$t = t^o/\sqrt{(1-v^2/c^2)}$

As the speed of the moving object increases to eventually reach light speed $v = c$ then $v^2/c^2 = 1$, and $\sqrt{(1-v^2/c^2)} = \sqrt{(1-1)} = 0$ so $t^o/0 = \infty$ (infinity). A consequence of dividing something by zero. This means the time between ticks becomes longer and longer until the time between ticks becomes infinite for the object now travelling at the speed of light. In other words, time there has effectively stopped - completely. It doesn't even exist. [79]

Objects travelling at light speed relative to ourselves, as do the 'gluons' holding us together, are effectively timeless. They exist outside of time, as does God Himself, *"from everlasting to everlasting."* (Psalm 90:2) Light itself has this massless, timeless existence as does God Himself. *"God is light"* (1 John 1:5) - so it says. Therefore, notwithstanding claims of Pantheism if it were not for Jesus, light is God - so it is. We ourselves are even referred to as *"children of light."* (Ephesians 5:8)

To take a concrete example in the real world of say a GPS satellite. This will reinforce the connection between scientific fact and biblical fact, as expressed in Einstein's universally accepted 'special theory of relativity', and God. Think of a GPS satellite in a stationary orbit above a point on the surface of the earth near where you and your mobile phone are both situated just now. It hardly seems like it, but the GPS satellite, the surface of the earth, you and your mobile phone are all rotating about a point at the centre of the earth.

Up there the satellite is further from this point of rotation at the centre of the earth than you are on the surface of the earth. The radius of its orbit is therefore longer. So, the GPS satellite must therefore travel further and therefore have a faster linear speed up there in order to keep tracking your smart phone down here.

Einstein's 'special theory of relativity' tells us, since the GPS satellite is travelling faster than you then time passes more slowly for it up there than for you down here, albeit ever so slightly.

To compensate for this out-of-synch situation, the atomic clock on the GPS satellite must be adjusted to run very slightly faster than the earth-bound clock linked to your smart phone. Without this adjustment they would quickly become unsynchronised and systematically miscalculate your position. Einstein's *"theory"* works in practise in this and many other applications.

Again to summarise, light always but always travels at the same speed regardless of the relative speeds of objects emitting and receiving the light. So, for any objects travelling at light speed time has not just stopped - it doesn't even exist. 'Gluons' are timeless and massless just like God, for He is spirit - as are we, our *supernatural* spirit selves inhabiting our *natural* bodies.

So to confirm, *"We know in part; then we shall know fully, as even now we are fully known."* (1 Corinthians 13:12) And yet, all is still not well with the world. What are we to make of all the things which can fall so drastically short of our expectations.

THE ONES LEFT BEHIND

Chapter 18

The Devil Was Defeated at the Cross - but How?

"In this world but not of this world"

John 8:23

To put some of this to the test of everyday life. Much of what goes badly in the world in the lives of those around us, and in our own lives, is accomplished by ourselves. We can mess things up pretty badly without any help from outside. However, things can happen which go so far beyond what any of us can believe to be an anywhere near normal part of *human nature*. Even we have limits.

The existence of disease, although not maybe its spread in every instance, is one example of what I will refer to as *evil* in the world. Premature death might be another, although again we as humans can contribute quite a lot to it all by ourselves. However, when premature death is brought about on a literally industrial scale along with so many individual acts of extreme malice wired into it, one does have to pause for at least a few moments, from time to time, to ask, *"Oh God!*

Oh God! Why?!" The Holocaust, the Holodomor in Ukraine and Armenian and Rwandan genocides come to mind, as well as certain individuals we hear about too frequently in the media.

Some say, *"You can't define evil.* When does one action cross the line from human nature to become evil? If you can't define the boundary line with evil then evil does not exist?" Well, neither can we define love, but few would say love does not exist, so I question the logic of the argument against evil.

Anyway, the embodiment of evil first makes his appearance in the Garden of Eden disguised as the serpent in Genesis 3:1. Long story short, not to give Satan any more space than to show how Jesus wins - end of - Jesus acknowledges Satan as *"the prince of this world."* (John 12:31, 14:30)

I think it's well to be assured, as we are when Jesus proclaims at the Last Supper, *"the prince of this world now stands condemned."* (John 16:11)

As Christians we all know the devil was defeated at the cross. How did it work? We accept this by faith, but I think non-Christians have a good question with, "Why is all this bad stuff happening? If God is love either he doesn't really care, or doesn't have the power to stop it..."

Of course, although we do have control of our lives, God is still in charge - of everything. He is Sovereign. So, I had assumed when Satan was defeated and condemned at the cross God just kind of pulled rank on him and strong-armed him out.

But being an engineer, and the person I am, it helps me to know more. I was intrigued by this word *"condemned"* referring to the *"prince of this world."* God loves justice which must be seen to be administered under the law. Satan could not be justly condemned simply because God is bigger

Chapter 18 - The Devil Was Defeated at the Cross

and stronger. Neither was there any jury tampering going on or anything.

What exactly was it about Jesus' death which condemned Satan under the law, for what crime and by which aspect of the law? The only person I ever heard explain this is David Yonggi Cho whose church I visited some years ago in Seoul, South Korea. At the time he was spinning off a new church of 100,000 members whenever his own core congregation reached 1,000,000 which I thought added credibility to his explanation, as well as it making Biblical sense.[80]

He put it this way. Satan was indeed condemned under the law in a court of law, the Court of the Heavenly Realms. Before any court can take on a legal case it must demonstrate it has jurisdiction over the accused. It must have the right, under the law of the land, to try the case. If a court does not have jurisdiction for whatever reason, it cannot bring a person to trial let alone convict and execute both justice and the convicted. If it does go ahead without having proper jurisdiction, then it in turn is in breach of the law of the land which does have jurisdiction.

God had taken Adam (presumably from the wild place where he'd created him) and, *"put him in the Garden of Eden to work it and take care of it."* (Genesis 2:15) God thereby handed Adam the keys to the Kingdom. Through his obedience to Satan Adam went on to give away his right of ownership of these keys, and thereby God's Kingdom, as in *"all authority and splendour"* to Satan. (Genesis 3:6 and Luke 4:5) Adam had obeyed Satan over eating the forbidden fruit instead of God by not eating it.

Back in the day, when Satan became the *"prince of this world"* he had legal jurisdiction over the world and everyone in it. He had authority in it and over it. As he declared to Jesus,

"...all the authority and splendour has been given to me, and I can give it to anyone I want to." (Luke 4:5) But how?

We empower the one we obey. This principle can work for or against us depending on who it is. If we give our agreement and obedience to the wrong authority we give them power over us. Adam empowered Satan over his life and over all those to come after him. He did this not by disobeying God, but by obeying Satan himself in eating the famous forbidden apple.

In this way, then Satan became, "the prince of this world" instead of Adam. (John 14:30) He had acquired legal jurisdiction over this world. He had authority in it and over it. As he declared to Jesus, "...all the authority and splendour has been given to me, and I can give it to anyone I want to." Echoed in (Luke 4:5) But how did that happen?

Although legally gifted to him by Adam under his authority, Satan went on to misappropriate the use of the keys by hugely oppressing the people and the land, the earth, to which they gave him access. Amongst the most important reasons Jesus had for coming here at all was to *"destroy the works of the devil."* (1 John 3:8) Amongst these reasons were specifically to recover the falsely acquired keys to the Kingdom and give them back to us, the "heirs of Adam." In this case, Jesus said, referring to the disciples, "I will give you the keys of the kingdom of heaven" (Matthew 16:19) but generally speaking giving them to "all those who believe." It remains now to us to re-appropriate those keys for ourselves into our daily lives under guidance from the Holy Spirit - the subject of many books written by people far more experienced than myself.

Long story short, Jesus was *tried* by the Sanhedrin, the Jewish religious authority of the day, for blasphemy, for claiming to be God. Their *guilty* verdict was punishable by death although they had no jurisdiction under the Roman law of occupation to exericse the penalty. Pilate however,

CHAPTER 18 - THE DEVIL WAS DEFEATED AT THE CROSS

in accordance with his jurisdiction as Roman Governor of Judea, was persuaded with great reluctance but under intense satanic pressure, to execute such a death sentence. Jesus was executed by crucifixion, as Satan had arranged and as foretold by God's curse in Genesis 3:15. Addressing all three in turn, Adam, Eve and Satan while referring to Jesus, God said, *"your offspring and hers (Jesus) will crush your (Satan's) head, and you (Satan) will strike his (Jesus') heel."*

Referring to just one of the gruesome aspects of Roman crucifixion, this comes with a health warning to some who may want to skip the rest of this paragraph and the reference. Nailed through the heels the body weight must also be supported by the hands, nailed to the cross-beam in order for the one being crucified simply to breathe. The legs are no longer able to assist the arms in taking breath. The agony of alternating one's bodyweight between nailed hands and nailed feet, can last for days. It leads inevitably to death through suffocation. The legs being unable to assist the hands any further makes it impossible to raise the body by the arms and hands enough to breath. Only breaking of the legs, so they no longer support the body in order to draw breath, brings the process to a faster *merciful* end.

However, although Jesus was in this world, he declared himself at his *"trial"* to be, *"not of this world."* (John 8:23) He was under the jurisdiction of God in the Kingdom of Heaven not of Satan on Earth. Satan was allowed, by design, to trick himself into thinking he had jurisdiction over Jesus because he was here when in fact he did NOT. Satan instead stood convicted of having arranged Jesus' extra-judicial death.

Of course, God knew how this would work out all along, since before the beginning of time. He knows the end from the beginning. (Isaiah 46:10) Jesus played his part by laying down his life in the manner he did as he knew he would, *"I lay down my life—only to take it up again."* (John 10:17) I like it that Satan didn't realise this and fell for the biggest *"sting"* of all time. The penalty forced him to return the keys to Jesus and he in turn to us. *"I will give you the keys of the kingdom of heaven; whatever you bind on earth will have been bound in heaven, and whatever you loose on earth will have been loosed in heaven."* (Matthew 16:9)

I also find this explanation quite satisfying and reassuring. It wasn't a matter of God being outraged and just pulling rank. As the God of Justice he acted within the law to condemn Satan after he himself over-played his hand. By illegally condemning Jesus to death outside his own jurisdiction he connived in his own condemnation and downfall.

Which takes me back to C. S. Lewis that, "The devil is a defeated enemy just as Hitler was defeated when the Allies secured the Normandy beachheads." Nevertheless, there had been much to fight for before they reached the inevitable conclusion in Berlin. Likewise for Charmaine, though the victory had been won from the outset there was much to fight for before she went home to be with Jesus and to receive the fullness of her healing.

Chapter 19

Conclusion

*"The most incomprehensible thing about
the Universe is that it is comprehensible."*
Albert Einstein

While avoiding the "God thing," with its understandable ramifications into Religion, Science maintains, "Something cannot come from nothing." Science is put to work uniquely in the domain of space, time and matter. Where they do not exist is defined as "nothing" - from which "something" cannot come. Catch 22, we shouldn't even be here - but we are. We know there are trillions of galaxies, each with an average 100 billion stars. The nearest is 2.5 million light years from here and most are billions of light years away. What we see happened hundreds of millions, if not billions of years ago. They may not even be there any more......

We as individuals are held together by the 'strong force' of God's love and massless 'gluon' particles traveling at light speed, which are timeless like God. They hold together every atom of our bodies and minds. And yet, all this is merely 5% of all there is in existence. The remaining 95% of 'dark matter' and 'dark energy' co-exist with and within us as the "heavenly realms."

And yet, our knowledge of What? is quite limited. To have any scientific knowledge of How? has been declared by Science to be beyond the realm of Science since the day of Galileo. Into the realm of "nothing measurable" it will not/can not go. Just how much bigger is the universe than our experience of church and traditional religion?

In this context, who are we to be so sure, based on what evidence, that we are not known and loved by the One who created it all? How can we be so certain that one day we will not "know fully as we are fully known?" (1 Corinthians 13:12)

> *"The important thing is not to stop questioning. Curiosity has its own reason for existence. One cannot help but be in awe when we contemplate the mysteries of eternity, of life, of the marvellous structure of reality. It is enough if one tries merely to comprehend a little of this mystery each day."*
> *Albert Einstein*

> *"For I am the Lord your God who takes hold of your right hand and says to you, 'Do not fear; I will help you.'"*
> *(Isaiah 41:13)*

Epilogue

The Way We Were

The first time, ever I saw your face
I thought the sun rose in your eyes
And the moon and the stars
Were the gifts you gave
And the first time, ever I lay with you
I knew our joy
Would fill the earth, my love
And last, till the end of time

Roberta Flack, "The First Time"[81]

In closing, I haven't shared Charmaine's last words to me, nor shall I. But I believe her first words to Jesus as he returned for her were as spoken here in this song by Amanda Cook:[82]

"You came, I knew that You would come.
I heard you call my name.
You sang, my heart it woke up.
I'm not afraid, I see Your face, I am alive!
You came, I knew that You would come.

THE ONES LEFT BEHIND

You said death's only sleeping. With one word my heart was beating. I rose up from my grave, my fear was turned to faith. Here in the waiting. I knew that You would come!!"

Yes, **I KNOW** she knew that He would come.

Glossary Of Terms

I have included a glossary as I am aware many people may not have a scientific background, especially one with huge cosmological emphasis (I include myself in this group). I hope these terms support your understanding of the themes and ideas I wish to convey in my book.

Atom: A particle consisting of a nucleus of protons and generally neutrons, surrounded by an electromagnetically-bound swarm of electrons. The atom is the basic particle of the chemical elements distinguished from each other by the number of protons in their atoms. For example, any atom that contains 11 protons is sodium, and any atom that contains 29 protons is copper. Atoms are extremely small. A human hair is about a million carbon atoms wide.

Big Bang: A theory describing how the universe expanded from an initial state of high density and temperature, otherwise referred to by Stephen Hawkins as "zero size and infinite density." These and other models offer a comprehensive explanation for a broad range of observed phenomena. A wide range of empirical evidence strongly favours the Big Bang event, which is now essentially universally accepted. Detailed measurements of the expansion rate of the universe, when reversed, place the Big Bang singularity back to an estimated 13.787±0.020 billion years ago. This is considered to be the age of the universe.

Conformal Cyclic Cosmology: A mathematical model in the framework of general relativity proposing a universe which iterates through infinite cycles. Each previous iteration is identified with a new Big Bang singularity representing one Big Bang followed by another infinite future expansion. Roger Penrose popularized this theory in his 2010 book Cycles of Time: An Extraordinary New View of the Universe. The hypothesis requires that all particles having mass eventually vanish from existence. Proton decay is a possibility contemplated in various speculative extensions of the Standard Model, but it has never been observed. Moreover, all electrons would also have to decay, or lose their charge and/or mass, and no conventional speculations allow for this.

Cosmic Inflation Theory: A theory of exponential expansion of space in the early universe. The inflationary epoch is believed to have lasted from 10^{-36} seconds to between 10^{-33} and 10^{-32} seconds after the Big Bang. Following the inflationary period, the universe still continued to expand from full size, but at a slower rate. The acceleration of this expansion due to dark energy began after the universe was already over 7.7 billion years old when the expansionary force of dark energy overtook that of gravity.

Cosmic Microwave Background (CMB): Microwave radiation that fills all space in the observable universe. Its faint background glow is almost completely uniform and is not associated with any star, galaxy, or other object. CMB is

GLOSSARY OF TERMS

landmark evidence of the Big Bang theory for the origin of the universe.

Dark Energy: An unknown form of energy that affects the universe on the largest scales. Its primary effect is to drive the accelerating expansion of the universe against the pull of gravity. Assuming that the lambda-CDM model of cosmology is correct, dark energy is the dominant component of the universe, contributing 68% of the total energy in the present-day observable universe. Dark energy's density is very low, much less than the density of ordinary matter or dark matter within galaxies. However, it dominates the universe's mass-energy content because it is uniform right across space.

Dark Matter: An unknown form of matter that appears to interact only with gravity but does not interact with light or the electromagnetic field. Dark matter is implied by gravitational effects which cannot be explained by general relativity unless more matter is present than can be seen. In the standard model of cosmology, the mass-energy content of the universe is 5% ordinary matter, 26.8% dark matter, and 68.2% a form of energy known as dark energy. Dark matter constitutes 85% of the total mass of the universe, while dark energy and dark matter together constitute 95% of its total mass-energy content.

Electron: A subatomic particle with a negative electric charge. Electrons are generally thought to be elementary particles because they have no known components or substructure. Electrons play an essential role in numerous physical phenomena, such as electricity, magnetism, chemistry, and thermal conductivity. Electrons are not themselves fundamental objects. Instead, there is spread throughout the entire universe something called an electron field, exactly like the quark and magnetic fields. The particles that we call electrons are little ripples of this electron field.

Element: A chemical substance that cannot be broken down into other substances. The basic particle that constitutes a chemical element is the atom, and each chemical element is distinguished by the number of protons in the nuclei of its atoms, known as its atomic number. For example, Oxygen has an atomic number of 8, meaning that each Oxygen atom has 8 protons in its nucleus.

Fine Tuned Universe: The characterization of the universe as finely tuned suggests that the occurrence of life in the universe is very sensitive to the values of certain fundamental physical constants. If the values of any of these free parameters in contemporary physical theories had differed only slightly from those observed, the evolution of the universe would have proceeded very differently, and the conditions for Life might not have been possible.

As Stephen Hawking has noted, *'The laws of science, as we know them at present, contain many fundamental numbers, like the size of the electric charge of the electron and the ratio of the masses of the proton and the electron. ... The remarkable fact is that the values of these numbers seem to have been very finely adjusted to make possible the development of life.'*

If, for example, the strong nuclear force was 2% stronger than it is while the other constants were left unchanged it is likely that all the universe's hydrogen would have been consumed in the first few minutes after the Big Bang.

Gluon: The so-called messenger particle of the strong nuclear force, which binds subatomic particles known as quarks within the protons and neutrons of stable matter. Quarks interact by emitting and absorbing gluons, just as electrically charged particles interact through the emission and absorption of photons of light.

GLOSSARY OF TERMS

Grand Unification Epoch: The period in the evolution of the early universe following the Planck epoch, starting at about 10^{-43} seconds after the Big Bang, in which the temperature of the universe was higher than 10^{27} degrees. During this period, three of the four fundamental forces - electromagnetic, strong force, and weak force but excluding gravity - were still unified in the one 'electronuclear force.'

This grand unification epoch ended at approximately 10^{-36} seconds after the Big Bang, at which point several key events took place. The strong force separated from the other fundamental forces. This phase transition is also thought to have triggered the process of cosmic inflation that dominated the development of the universe during the following inflationary epoch.

Grand Unified Theory (GUT): is any model in particle physics that merges the electromagnetic, weak, and strong forces into a single force at high energies. Although this unified force has not been directly observed, many GUT models theorize its existence. If the unification of these three interactions is possible, it raises the possibility that there was a grand unification epoch in the very early universe in which these three fundamental interactions were not yet separated and distinct.

Gravity: Causes mutual attraction between all things that have mass. Gravity is, by far, the weakest of the four fundamental forces, approximately 10^{38} times weaker than the strong force, 10^{36} times weaker than the electromagnetic force and 10^{29} times weaker than the weak force. As a result, it has no significant influence at the level of subatomic particles at the micro-cosmic level. However, gravity is the most significant interaction between objects at the cosmic scale, as it determines the motion of planets, stars, galaxies, and even light.

On Earth, gravity interacts with the mass of matter giving weight to physical objects. Gravity has an infinite range, although its effects become weaker as objects get farther away. The gravitational attraction between the original gaseous matter in the universe caused it to coalesce and form stars which eventually grouped together into galaxies.

For most applications, gravity is well approximated by Newton's law of universal gravitation, which describes gravity as a force causing any two bodies to be attracted toward each other, with magnitude proportional to the product of their masses and inversely proportional to the square of the distance between them.

Higgs Field: An energy field thought to exist in every region of the universe. The Higgs field is accompanied by a fundamental particle known as the Higgs boson. This is used by the Higgs field to continuously interact with other particles thereby giving them mass and in turn weight when interacting with gravity. If the Higgs field did not exist, particles would not have the mass required for gravity to attract them to one another. They would float around freely - at light speed.

Light Year: A unit of length used to express astronomical distances equal to approximately 5.88 trillion miles. Because it includes the word 'year' the term is sometimes misinterpreted as a unit of time. A light year is the distance light travels in a year at light speed, namely 186,000 miles second.

Milky Way: The galaxy that includes our solar system, with the name describing the galaxy's appearance from Earth: a hazy band of light seen in the night sky formed from stars that cannot be individually distinguished by the naked eye.

The Milky Way is a barred spiral galaxy with a diameter estimated at 87.40 light years but only about 1,000 light-years thick at the spiral arms (more at the bulge). Recent

GLOSSARY OF TERMS

simulations suggest that a dark matter area, also containing some visible stars, may extend up to a diameter of almost 2 million light-years.

It is estimated to contain 100-400 billion stars and at least that number of planets. The constant rotational speed appears to contradict the laws of Keplerian dynamics and suggests that much (about 90%) of the mass of the Milky Way is invisible to telescopes, neither emitting nor absorbing light or any other kind of electromagnetic radiation. This conjectural mass has been termed "dark matter."

The rotational period is about 212 million years at the radius of the Sun. The Milky Way as a whole is moving at a velocity of approximately 600 km per second (372 miles per second) with respect to extragalactic frames of reference. The oldest stars in the Milky Way are nearly as old as the Universe itself.

Neutron: a subatomic particle made from two down quarks and one up quark. It has a neutral (not positive or negative) charge, and a mass slightly greater than that of a proton. Protons and neutrons constitute together the nuclei of atoms.

Neutron Star: The collapsed core of a massive supergiant star, which had a total mass of between 10 and 25 solar masses Neutron stars have a radius on the order of 10 kilometres. Due to the extreme gravitational pressure, the electrons and protons present in normal matter combine to produce more neutrons so that neutrons alone remain.

Neutron star material is remarkably dense: a normal-sized matchbox containing neutron-star material would have a weight of approximately 3 billion tonnes, the same weight as a 0.5-cubic-kilometer chunk of the Earth, a cube with edges of about 800 meters. As a star's core collapses, its rotation rate increases due to conservation of angular momentum. The fastest-spinning neutron star known is PSR J1748-2446ad, rotating at a rate of 716 times per second giving a linear speed at the surface on the order of nearly a quarter the speed of light. There are thought to be around one billion neutron stars in the Milky Way.

Nuclear Fission: A reaction in which the nucleus of an atom splits into two or more smaller nuclei. The fission process often produces gamma photons and releases a very large amount of energy. Liberation of additional neutrons during the fission process opens up the possibility of a nuclear chain reaction.

Nuclear Fusion: When two small, light nuclei join together to make one heavy nucleus. Fusion reactions occur in stars where two hydrogen nuclei fuse together under high temperatures and pressure to form a nucleus of helium. Nuclear fusion is the principal which drives the Hydrogen bomb.

Nucleus: the small, dense region consisting of protons and neutrons at the centre of an atom. Almost all the mass of an atom is located in the nucleus, with a very small contribution from the electron cloud. Protons, neutrons and electrons are bound together to form a nucleus by the strong and the electromagnetic forces.

Planck Epoch: In Big Bang cosmology, the Planck epoch or Planck era is the earliest stage of the Big Bang, before the time passed which was equal to the Planck time of approximately 10^{-43} seconds. There is no currently available physical theory to describe what happens in such short times, and it is not clear in what sense the concept of time is meaningful for values smaller than the Planck time. At this scale, the unified force of the Standard Model is assumed to be unified with gravitation. Immeasurably hot and dense, the state of the Planck epoch was succeeded by the grand unification epoch, where gravitation separated from the unified force of the

GLOSSARY OF TERMS

Standard Model, in turn followed by the inflationary epoch, which ended after about 10^{-32} seconds.

Proton: a stable subatomic particle made from two up quarks and one down quark. Its mass is slightly less than that of a neutron and 1,836 times the mass of an electron. One or more protons are present in the nucleus of every atom. The number of protons in the nucleus is the defining property of an element and is referred to as the atomic number. Since each element has a unique number of protons, each element has its own unique atomic number, which determines the number of electrons in each atom and consequently the chemical characteristics of the element.

Quantum Mechanics: General Relativity is the best theory of the "very big" - the cosmic. 'Quantum Mechanics' is the best theory of the "very small" - the micro-cosmic, of which everything is made. Take a simple example of money. Money is measured in small quantities or quanta. A quantum of money is a penny or a cent depending on the currency. It's a clearly defined object or entity, the least amount of that currency which can exist. It cannot exist in any smaller quantities. Cut a penny in half and it no longer exists as money. We can come across £1 coins or $1 bills of course. They represent 100 pence, 100 cents or 100 quanta of money, but they are "made of" pennies and cents.

Likewise light, energy and even time occur in quanta. They are minimum quantities that can be no less without ceasing to exist as such at all. Quanta of light are known as 'photons' of a specific frequency. Electrons are the smallest quanta of electrical energy. Even time does not exist in the continuous flow it appears to be. Time too exists in quantum bits of no less than 10^{-43} seconds, like the flickering frames of a celluloid movie that seem continuous when viewed fast enough one after the other. This quantum of time is the closest we can get in time to the Big Bang event itself because this Planck time of 10^{-43} seconds cannot be divided into something smaller. Within this so-called 'Planck Epoch' there is - no time. Time simply does not exist, just as money does not exist in values of less than a penny or a cent.

Quark: A type of elementary particle produced by ripples in the Quark Field and a fundamental constituent of matter. Quarks combine to form composite particles, the most stable of which are protons and neutrons, the components of atomic nuclei. All commonly observable matter is made of 'up quarks' 'down quarks' and electrons.

Quark Field: Quarks are not themselves fundamental objects. Instead, spread throughout the entire universe is something called a quark field, exactly like the electric and magnetic fields. And the particles that we call quarks are tiny eddies or ripples of this quark field.

Relativity - General: Einstein's theory published in 1915 is the current description of gravitation in modern physics. General Relativity provides a unified description of gravity as a geometric property of space and time or four-dimensional spacetime which becomes 'curved' in the presence of mass. The greater the mass, the greater the curvature of spacetime and with it the curvature of a path of light passing through it.

Such mass can become so great in the instance of a 'black hole,' the curvature of spacetime so intense, light curves back on itself and cannot leave from it.

Relativity - Special: The special theory of relativity, is a scientific theory of the relationship between space and time. In Albert Einstein's 1905 treatment, the theory is presented as being based on just two postulates:
1. The laws of physics are identical in all frames of reference with no acceleration.
2. The speed of light in a vacuum is the same for all observers, regardless of the motion of light source or observer.

GLOSSARY OF TERMS

Combined with other laws of physics, the two postulates of special relativity predict the equivalence of mass (m) and energy (E.) One can become the other in proportions expressed in Einstein's mass-energy equivalence formula $E=mc^2$ where c is the speed of light in a vacuum.

Spacetime: The four-dimensional geometry of up, right, sideways and time. Gravity is not simply an external force that acts on massive bodies, as viewed by Isaac Newton's universal gravitation. Instead, general relativity links gravity to spacetime itself which is not 'flat' but is curved by the presence of massive bodies. The curvature of spacetime influences the motion of bodies having mass within it. In turn, as massive bodies move in spacetime its curvature changes. The geometry of spacetime is in constant evolution. Gravity is a result of the dynamic interaction between matter and spacetime.

Standard Model: The theory in particle physics describing three of the four known fundamental forces in the universe (electromagnetic, weak and strong interactions - excluding gravity) and classifying all known elementary particles. It was developed in stages throughout the latter half of the 20th century, through the work of many scientists worldwide with the current formulation being finalized in the mid-1970s upon experimental confirmation of the existence of quarks.

String Theory: A theoretical framework in which the point-like particles of particle physics are replaced by one-dimensional objects called strings. String theory describes how these strings propagate through space and interact with each other. On distance scales larger than the string scale, a string looks just like an ordinary particle, with its mass, charge, and other properties determined by the vibrational state of the string. In string theory, one of the many vibrational states of the string corresponds to the graviton, a quantum mechanical particle that carries the gravitational force. Thus, string theory is a theory of quantum gravity. Despite much work on these problems, it is not known to what extent string theory describes the real world or how much freedom the theory allows in the choice of its details.

Strong Force: Holds together the building blocks of atoms. It always attracts and works at two different size scales in atoms. At the level of an atomic nucleus, the strong force holds together the protons and neutrons that form the essence of the elements.

As suggested by its name, the strong force is the strongest of the fundamental forces. It is about 100 times stronger than electromagnetism and 100 trillion trillion trillion times stronger than gravity.

However, the strong force only has influence over very, very small distances. For anything larger than the nucleus of a medium-sized atom (about 100 million times smaller than the width of a human hair), its influence quickly drops, and other forces will be stronger.

Supernova: A powerful and luminous explosion of a star. A supernova occurs during the last evolutionary stages of a massive star or when a white dwarf is triggered into runaway nuclear fusion. The original object, called the progenitor, either collapses to a neutron star or black hole, or is completely destroyed to form a diffuse nebula. The peak optical luminosity of a supernova can be comparable to that of an entire galaxy before fading over several weeks or months.

Time: The continued sequence of existence and events that occurs in an apparently irreversible succession from the past, through the present, and into the future. Time is often referred to as a fourth dimension, along with three spatial dimensions - up, down, sideways.

References and Endnotes

When navigating through the pages of a book, where information has been referenced can create a source of frustration, regardless of whether the reference pertains to a book, a piece of music, websites or another source. The challenge lies in deciphering which reference corresponds to which section. This can often be difficult to locate and explore cited material further. To alleviate this frustration, I have adopted a comprehensive approach by presenting all references in a list numbered sequentially across the chapters. I hope this provides you with a seamless experience, enabling you to effortlessly delve into the content rich sources which inspired me.

THE ONES LEFT BEHIND

Preface

1 **Di Marco, Kristine,** *It Is Well,* YouTube, https://www.youtube.com/watch?v=TodIWJ4t4Jg 22/06/2023

Chapter 1

2 **Clegg, Brian,** *Dark Matter Dark Energy,* Good Reads, https://www.goodreads.com/book/show/44784403-dark-matter-and-dark-energy 22/06/2023

3 **Wallace, David Foster,** *This is Water,* Wikipedia https://en.wikipedia.org/wiki/This_Is_Water 22/06/2023

4 **Arvin, Ash,** *General Relativity explained simply,* YouTube, https://www.youtube.com/watch?v=tzQC3uYL67U 22/06/2023

5 **Hawking, Stephen,** *A Brief History of Time,* Wikipedia, https://en.wikipedia.org/wiki/A_Brief_History_of_Time 22/06/2023

6 **Editors, The,** *Dark Energy and Dark Matter,* Wikipedia https://en.wikipedia.org/wiki/Dark_energy https://en.wikipedia.org/wiki/Dark_matter 26/06/2023

7 **Morris, Desmond,** *The Naked Ape,* https://en.wikipedia.org/wiki/The_Naked_Ape 22/06/2023

8 **Editors, The,** *Neutron Star,* Wikipedia, https://en.wikipedia.org/wiki/Neutron_star 22/06/2023

9 **Tong, David,** *Quantum Fields: The Real Building Blocks of the Universe,* YouTube https://youtu.be/zNVQfWC_evg 23/04/2024

10 **Editors, The,** *Special Relativity,* Wikipedia, https://en.wikipedia.org/wiki/Special_relativity 22/06/2023

11 **Wisdom for Life,** *The Living Universe is Unimaginably BIG and You're a Part of it!,* youtube https://youtu.be/xVLZ8_AzD1c 23/04/2024

12 **Fermilab,** *Photons,* YouTube, https://youtu.be/6Zspu7ziA8Y 22/06/2023

13 **St Augustine,** *On Temporality and Consciousness,* AZ Quotes, https://www.azquotes.com/author/663-Saint_Augustine/tag/time 22/06/2023

REFERENCES

14 **Editors, The**, *Conformal Cyclic Cosmology*, Wikipedia, https://en.wikipedia.org/wiki/Conformal_cyclic_cosmology 22/06/2023

15 **Editors, The**, *Pheromones*, Wikipedia, https://en.wikipedia.org/wiki/Pheromone 22/06/2023

Chapter 2

16 **Heiligenthal, Kalley**, *Shepherd*, YouTube, https://youtu.be/V7vaFvqVTJ8 22/06/2023

Chapter 3

17 **Backlund, Steve**, *You're Crazy if You Don't Talk to Yourself*, Good Reads, https://www.goodreads.com/book/show/7773348-you-re-crazy-if-you-don-t-talk-to-yourself?from_search=true&from_srp=true&qid=3ui0Bhx021&rank=8 22/06/2023

18 **Heiligenthal, Kalley**, *You Know Me*, YouTube, https://youtu.be/RLpGFULSk8Q 22/06/2023

19 **Chambers, Oswald**, *My Utmost for His Highest - August 26th Providential Permission*, Oswald Chambers, https://www.oswaldchambers.co.uk/classic/are-you-ever-disturbed-classic/ 22/06/2023

20 **Webb, Richard**, *The Strong Force*, New Scientist, https://www.newscientist.com/definition/strong-nuclear-force/ 23/04/2024

21 **Di Marco, Kristene**, *Lilly's Song*, YouTube, https://youtu.be/1kFEgMYJyZo 26/06/2023

Chapter 6

22 **Gaga, Lady**, *Million Reasons*, YouTube, https://youtu.be/JezHPWOpPtc 22/06/2023

23 **Smith, Martin**, *Song of Solomon*, YouTube, https://youtu.be/WTXCo88EAao 23/04/2024

24 **Jesus Culture**, *Dance with Me*, YouTube, https://youtu.be/sSbVGbQcLAQ 22/06/2023

25 **Di Marco, Kristene**, *It is Well*, YouTube, https://youtu.be/TodIWJ4t4Jg 22/06/2023

26 **Editors, The**, *Euclid Spacecraft*, Wikipedia, https://en.wikipedia.org/wiki/Euclid_(spacecraft) 22/06/2023

Chapter 7

27 **Editors, The**, *Ffald-y-Brenin*, Ffald-y-Brenin, https://ffald-y-brenin.org/ 22/06/2023

28 **Hatch, Edwin**, *Breathe on me Breath of God*, YouTube, https://youtu.be/MmkzSjs9eAw 22/06/2023

Chapter 9

29 **Maguire, Mark John**, *Zulu*, YouTube, https://youtu.be/HUq8gXhIoy8 22/06/2023

Chapter 10

30 **Editors, The**, *Lewis, C. S.: Aslan*, Wikipedia, https://en.wikipedia.org/wiki/Aslan 22/06/2023

31 **Walker-Smith, Kalley**, *Pursue*, YouTube, https://youtu.be/SkPe7jYWh10 26/06/2023

32 **Editors, The**, *Cognitive Bias*, Wikipedia, https://en.wikipedia.org/wiki/Cognitive_bias 26/06/2023

33 **Madrugada Eterna**, *Alfie Final Scene*, YouTube, https://www.youtube.com/watch?v=B8166-kaEPQ 23/04/2024

34 **Hetland, Leif**, *Destined for the Throne*, Leif Hetland, https://www.leifhetland.com/destined-for-the-throne/ 22/06/2023

35 **Clegg, Brian**, *Dark Matter Dark Energy*, https://brianclegg.net/dmde.html 22/06/2023

36 **Editors, The**, *Steve Jobs: Final hours*, Wikipedia, https://en.wikipedia.org/wiki/Steve_Jobs#Death 22/06/2023

37 **Editors, The**, *Dudley Moore: Final hours*, Wikipedia, https://en.wikipedia.org/wiki/Dudley_Moore 22/06/2023

38 **Chambers, Oswald**, *The Warning Against Wantoning, April 24 Devotional: Oswald Chambers*, https://www.thespiritlife.net/65-devotional/devotional-publications/682-april-24-devotional-oswald-chambers

Chapter 11

39 **Tong, David**, *Physics - Atoms*, YouTube, https://youtu.be/VpyicsMIJU 22/06/2023

Chapter 12

40 **Editors, The**, *Black Holes*, Wikipedia, https://en.wikipedia.org/wiki/Black_hole 22/06/2023

REFERENCES

41 **Stevenson, Mary**, *Footsteps in the Sand*, https://www.onlythebible.com/Poems/Footprints-in-the-Sand-Poem.html 24/04/2024

42 **Editors, The**, *String Theory*, Wikipedia, (https://en.wikipedia.org/wiki/String_theory 22/06/2023

43 **Editors, The**, *Gluons*, Wikipedia, https://en.wikipedia.org/wiki/Gluon 22/06/2023

44 **Editors, The**, *Steve Jobs: Final hours*, Wikipedia, https://en.wikipedia.org/wiki/Steve_Jobs#Death 22/06/2023

45 **Editors, The**, *Dudley Moore: Final hours*, Wikipedia, https://en.wikipedia.org/wiki/Dudley_Moore 22/06/2023

46 **Editors, The**, **Planck Epoch**, Wikipedia, https://simple.wikipedia.org/wiki/Planck_epoch 22/06/2023

47 **Editors, The**, *Descartes, Rene, Cogito, ergo Sum*, Wikipedia, https://en.wikipedia.org/wiki/Cogito,_ergo_sum 22/06/2023

48 **Editors, The**, *Quarks*, Wikipedia, https://en.wikipedia.org/wiki/Quark 22/06/2023

49 **Osborne, Joan**, *What if God was One of Us*, YouTube, https://www.youtube.com/watch?v=aDdOnlobHO4 22/07/2023

50 **Editors, The**, *Enzymes*, Wikipedia, https://en.wikipedia.org/wiki/Enzyme 22/06/2023

Chapter 13

51 **Editors, The**, *Inflation (cosmology)*, Wikipedia, https://en.wikipedia.org/wiki/Inflation_(cosmology) 23/04/2024

52 **Editors, The**, *The Most Distant Astronomical Objects*, Wikipedia, https://en.wikipedia.org/wiki/List_of_the_most_distant_astronomical_objects 22/06/2023

53 **Smethurst, Dr Becky**, *How Big is the Entire Universe?*, YouTube, https://youtu.be/6kJoI7SyJsU 22/06/2023

54 **Editors, The**, *Planck Epoch - Chronology of the Universe*, Wikipedia, https://en.wikipedia.org/wiki/Chronology_of_the_universe 22/06/2023

55 **Editors, The**, *Fine-tuned Universes*, Wikipedia, https://en.wikipedia.org/wiki/Fine-tuned_universe 22/06/2023

56 **Editors, The**, *Anthropic Principle*, Wikipedia, https://en.wikipedia.org/wiki/Anthropic_principle 22/06/2023

57 **Editors, The**, *Atomic clock*, Wikipedia, https://en.wikipedia.org/wiki/Atomic_clock 22/06/2023

58 **Editors, The**, *Cosmic Microwave Background - CMB*, Wikipedia, https://en.wikipedia.org/wiki/Cosmic_microwave_background 22/06/2023

59 **Editors, The**, *The Cosmological Constant*, Wikipedia, https://en.wikipedia.org/wiki/Cosmological_constant 22/06/2023

60 **Editors, The**, *Multiverses*, Wikipedia, https://en.wikipedia.org/wiki/Multiverse 22/06/2023

61 **Suskind, Leonard**, *Is the Universe Fine-Tuned for Life and Mind?*, YouTube, https://youtu.be/2cT4zZIHR3s 22/06/2023

Chapter 14

62 **Editors, The**, *Ozone*, Ozonewatch, https://ozonewatch.gsfc.nasa.gov/facts/dobson_SH.html 22/06/2023

63 **Editors, The**, *Carbonaceous Chondrites*, Wikipedia, https://en.wikipedia.org/wiki/Carbonaceous_chondrite 22/06/2023

64 **Editors, The**, *Chlorophyll*, Wikipedia, https://en.wikipedia.org/wiki/Chlorophyll 22/06/2023

65 **Editors, The**, *Hobbes, Thomas, On Life Solitary, Poor, Nasty, Brutish and Short*, Yale books blog, https://yalebooksblog.co.uk/2013/04/05/thomas-hobbes-solitary-poor-nasty-brutish-and-short/ 22/06/2023

66 **Editors, The**, *Neutrinos*, Wikipedia, https://en.wikipedia.org/wiki/Neutrino 22/06/2023

67 **Editors, The**, *Strong interaction*, Wikipedia, https://en.wikipedia.org/wiki/Strong_interaction 23/04/2024

67a **Walker-Smith, Kim**, *Heaven is Here Now*, YouTube, https://youtu.be/w6pGtBuHOYA 28/06/2024

68 **apbiolghs**, *Gravity Visualised*, YouTube, https://youtu.be/MTY1Kje0yLg 22/06/2023

69 **SpaceSciNewsroom**, *Gravity Demo Part 2 Basic DemoGravity Visualised*, YouTube, https://www.youtube.com/watch?v=MkbUzfTq3w4 22/06/2023

REFERENCES

70 **Editors, The**, *Higgs Field*, Wikipedia, https://simple.wikipedia.org/wiki/Higgs_field#:~:text=The%20Higgs%20field%20is%20a%20field%20of%20energy,interact%20with%20other%20particles%2C%20such%20as%20the%20electron 22/06/2023

71 **Editors, The**, *Quark*, Wikipedia, https://en.wikipedia.org/wiki/Quark 22/06/2023

72 **Editors, The**, *Sagittarius A*, Wikipedia, https://en.wikipedia.org/wiki/Sagittarius_A* 22/06/2023

73 **Tong, David**, *Quantum Fields: The Real Building Blocks of the Universe*, The Royal Institution, YouTube, https://youtu.be/zNVQfWC_evg 22/06/2023

74 **Frankl, Viktor**, *Man's Search for Meaning*, Good Reads, https://www.goodreads.com/book/show/4069.Man_s_Search_for_Meaning 22/06/2023

Chapter 15

75 **Chambers, Oswald**, *My Best for His Highest - May 6th, The Spirit Life*, https://www.thespiritlife.net/65-devotional/devotional-publications/773-may-6-devotional-oswald-chambers 22/06/2023

Chapter 16

76 **Editors, The**, *Coma Cluster*, Wikipedia, https://en.wikipedia.org/wiki/Coma_Cluster 22/06/2023

77 **Di Marco, Kristine**, *Take Courage - He's in the Waiting*, YouTube, https://www.youtube.com/watch?v=r49V9QcYheQ 22/06/2023

Chapter 17

78 **Science ABC**, *Time Dilation - Einstein's Theory Of Relativity Explained!*, YouTube, https://youtu.be/yuD34tEpRFw 23/04/2024

79 **Physics Explained**, *Deriving Einstein's most famous equation: Why does energy = mass x speed of light squared?*, (E=mc2), YouTube, https://youtu.be/KZ8G4VKoSpQ 22/06/2023

Chapter 18

80 **Sullivan, David,** *Died: David Yonggi Cho, Founder of the World's Largest Megachurch,* Christianity Today, https://www.christianitytoday.com/news/2021/september/died-david-yonggi-cho-korea-megachurch-cell-church-growth.html 22/06/2023

Epilogue

81 **Flack, Roberta,** *First Time Ever I Saw Your Face 1972,* https://www.youtube.com/watch?v=d8_fLu2yrP4 23/04/2024 & **Editors, The,** *Roberta Flack Lyrics,* AZLyrics, https://www.azlyrics.com/lyrics/robertaflack/thefirsttimeeverisawyourface.html 26/04/2024

82 Cook, Amanda, Lazarus, YouTube, https://youtu.be/gRG5U-RaVdI 28/06/2024

Other References

Rilke, Rainer Maria, *Time and Again,* All Poetry, https://allpoetry.com/Time-and-Again 30/7/23

Acknowledgements

This book has been simmering in my mind for the last 10 years but was especially inspired by Benni Johnson and the team who accompanied me to Israel in May 2015. "You have to write a book!" they said. "Who me? Only people who write books do that and I never had before.

Its many moving parts began to arrive in my mind and were dictated into Notes on my iPhone. They focused first of all on the recent passing of my wife and my feeling like "the one left behind," which became the title for the first 45 page book, the foundation for the current one.

My mother's passing years before, some years after my father's, had brought about a spiritual crisis which evangelist Melvin Banks suggested I address by reading the Bible every day. "Oh No! I thought. Has it really come to this?!" But the good news he assured me was that I didn't need to understand it. I proved to be quite good at that - for years. Nevertheless, God finally allowed multiple circumstances to lead us to Cornerstone Christian Fellowship in France. There pastor Alan Valentine, church members James and Jill Mills, Ed and Sam Cartwright, Daniel and Danielle Angevert, Elie and Judith Ferraro and many others, helped bring things into focus for us through the reality that Jesus is alive and had been there all our lives.

Later, having been impacted by Bill Johnson and all things Bethel Church, Redding, California his book made clear to us "On Earth as it is in Heaven" is for now and that God is Good all the time.

After Charmaine's passing my father's lifelong fascination with all things scientific, his awesome wonder at creation, then began to kick in. Scientific factoids began to accumulate and match themselves to Bible verses in ways I'd never known before. Ladey Adey mid-wifed all this for me into this book. She inspired its current title, walked through the writing of each chapter with me. Together with her daughter Abbirose we hammered out the boiler-plate, so to speak.

Meanwhile Andrew and Sarah Kampouris, Rosie Park, Julie Tushingham, Ben and Helen Wright, Monica Brown, Andy and Ali Shepherd, Martin and Anne Wallis and Julie Thompson have partnered with me over recent years, challenging, supporting and encouraging me to keep going.

To all the above, and many more than can be mentioned, each for their unique contributions, goes my gratitude.

About the Author

Mike's working life spans multiple continents, and multiple careers in sales and marketing and manufacturing consulting.

Finding new and innovative things, not in what nobody else has seen but in what everybody else has seen, has been a source of his added value.

Coming late to Faith he brings an 'outsider's view' to this book. Aligning science fact with biblical fact he journeys through love and loss, through faith and hope with those who have lost a loved one and feel left behind.

Mike and his wife Charmaine met on their first day at university in Leeds where he obtained an Honours Degree in Mechanical Engineering and she in English and Arabic. They lived and worked together in cities across Europe and the Middle East. Their daughter is bilingual in English and French. Having been brought up in Paris and Nice she now

THE ONES LEFT BEHIND

lives with her husband and four children near Vancouver where Mike has recently joined them and is looking forward to discovering the wonders of Canada.

Index

Symbols
π 130, 163

A
Adey, Abbirose iv
Aeschylus 23
Alighieri, Dante 101
Alpha Centauri 14
Andromeda 14
Archimedes 132
Aristotle 120, 132
Attenborough, David 28, 113

B
Backlund, Steve 39
Bethel Church 46, 47, 79
Bethel Music 29, 63, 86, 89
Big Bang 2, 4, 13, 16, 18, 20, 21, 22, 127, 133, 134, 137, 138, 149, 151, 152, 157, 158, 160, 161, 168, 169, 171, 172, 196, 197, 211, 217, 225
Big Birth 4, 133, 177
Big Crunch 168, 217
Bishop of London xi
Blake, Willian 121
Bohr, Niels 132
Bruce Almighty 20

C
Caesium Clock 165
Chambers, Oswald 89, 113, 204
Chesapeake Bay 50, 51
Cognitive Bias 96
Cognitive Dissonance 15
Coma Galaxy Cluster 215
Conformal Cyclic Cosmology 22
Cosmic Microwave Background 167
Cosmological Constant 168, 169
Cosmological Inflation Theory 159
Cosmological Microwave Background 134
Cox, Professor Brian 113

D
Dark Energy 10, 71, 107, 122, 131, 137, 141, 168, 169, 170, 190, 191, 196, 197, 207, 212, 213, 214, 216, 217, 218, 240
Dark Matter 7, 10, 71, 107, 122, 127, 131, 137, 141, 170, 190, 191, 196, 197, 207, 212, 213, 214, 215, 216, 217, 218, 220, 221, 222, 240
Dawkins, Richard 22, 132
Descartes, Rene 136, 182, 185

E
Einstein, Albert 4, 13, 16, 21, 22, 128, 132, 153, 158, 159, 168, 173, 189, 196, 198, 201, 226, 230, 231, 240
Electron Field 12
Eternal Life 18, 45, 98, 109, 115, 116, 146, 168, 220, 226
Euclid Telescope 10, 71

F
Ffald-y-Brenin 80
Flack, Roberta xi, 241
Footsteps in the Sand 129
Frankl, Viktor 201
Freeman, Morgan 19

G
Galileo, Galilei 130, 132, 240
Gell-Mann, Murray 194
General Relativity 8
Gluons 131, 196, 201, 228, 230, 231, 239
Grace 23, 46, 47, 52, 79, 92, 112, 113, 137
Grand Unified Theory (GUT) 9
Gravity 10, 117, 119, 120, 158, 159, 168, 169, 173, 176, 180, 181, 182, 188, 189, 190, 191, 192, 193, 196, 197, 198, 199, 200, 205, 212, 214, 217

H
Hawking, Stephen 9
Hawkins, Stephen 243, 246
Heisenberg, Werner 132
Hetland, Leef 107
Hickman Line 52
Higgs Boson 197, 198, 200, 218
Higgs Field 12, 117, 191, 197, 198, 200, 201, 218
Higgs, Peter 198

267

Hippocrates 132
Hobbes, Thomas 179
Holy Spirit 34, 38, 52, 72, 76, 83, 90, 95, 151, 178, 187, 207, 218, 236
Hubble, Edwin 4, 132, 168

J

JADES-GS-z13-0 160
James Webb Telescope 158
Jesus Culture 68
Jobs, Steve 107, 132
Johnson, Bill 46, 62, 79
Joyce, James 171, 194
    Finnegan's Wake 171

K

Kavil Prize 159
Kelvin, Lord 215

L

Lady Gaga 64
Lewis, C. S. 40, 94, 105, 143, 164, 238
    Beyond Personality 143, 145, 179

M

Mainwaring, Paul 47, 53
Milky Way 14, 15, 122, 175, 200, 215
Moore, Dudley 108, 132
Morrison, Jim 144
Mother Teresa 112
Mount Everest 154
M Theory 21

N

NASA 14
Nasser, Colonel Gamal Abdel 53
Newton, Isaac 4, 132, 192

O

Oromorph 56, 57, 58
Osborne, Joan 19, 145

P

Pascal, Blaise 139, 218
Pasternak, Boris xi
Penrose, Prof. Roger 22
Penzias, Arno 167
Periodic Table 126
Planck Epoch 134, 149, 151, 160
Planck, Max 134
Plato 136
Probabilistic Cloud Layers 192
Protagoras 17
Proxima Centauri 14

Q

Quantum Mechanics 8, 115
Quark Field 12
Quarks 144, 151, 166, 171, 188, 191, 194, 195, 196, 197, 198, 200, 201, 228

R

Rorke's Drift 91
Rutherford, Ernest 4

S

Sagittarius A 200
Shakespeare, William
    Hamlet 16, 125, 175, 179
    Macbeth 18
Simpson, Mona 107
Smith, Martin 65
Spacetime 4, 21, 22, 139, 157, 189, 190, 191, 195, 196, 198, 199, 200, 214, 226
St Augustine 20
String Theory 21, 129
Strong Force 45, 115, 117, 119, 120, 131, 144, 151, 165, 176, 181, 182, 201, 239
Suez Canal 53

T

ten Boom, Corrie 95
Time Dilation 229
Tong, David 201

V

Voyager 1 14

W

Washington, Denzel 27, 116, 120
Williams, Rowan 205
Wilson, Robert 167
Wordsworth, William 183, 213

Z

Zulu 91, 201

Notes

You may find it useful to note down how you feel when reading this book. If something resonates deep within you or even you just want a quick reminder of how you feel whilst reading a certain chapter.

_____
_____
_____
_____
_____
_____
_____
_____
_____

THE ONES LEFT BEHIND

NOTES

THE ONES LEFT BEHIND

NOTES

THE ONES LEFT BEHIND

NOTES

THE ONES LEFT BEHIND

NOTES

THE ONES LEFT BEHIND

NOTES

THE ONES LEFT BEHIND

www.ingramcontent.com/pod-product-compliance
Lightning Source LLC
Chambersburg PA
CBHW040240130526
44590CB00049B/4032